EMP Hoax

EMP Hoax

David Hathaway

CharityEndureth.com

David Hathaway
441 N. Grand Avenue, Suite 4
Nogales, Arizona 85621
CharityEndureth.com
Author email: charityendurethallthings@gmail.com

ISBN-13: 978-1983751752

To my grandson Lukas

Contents

Foreword

Scholars like Garet Garrett, John T. Flynn, and Murray Rothbard have pointed out that a main way the State adds to its power and tyrannizes over the people is to whip up alleged threats from a foreign power. Given the threat, defenders of the State say, don't we need to take countermeasures, however high the cost? Those of us who grew up in the '50s and '60s remember vividly the "drop drills" that prepared schoolchildren for an impending nuclear attack. The government neglected to tell us that America was far better armed than the Soviets. The "threat" was concocted to justify a militarist policy and massive government spending.

In this careful study, David Hathaway has given us a microstudy of the Orwellian State's technique to add to its power. On July 8, 1962, an atomic warhead was detonated in the Pacific. One purpose of the blast was to study the impact, if any, of ElectroMagnetic Pulse (EMP) effects. One incident is alleged to show such effects. Based on this incident, the government concluded that hostile powers could use EMPs to disable the electronic infrastructure of our country. Even after the Cold War ended, the government has continued to tout the threat posed by EMPs.

Hathaway comments: "The alleged incident happened on the island of Oahu which is made up of the City and County of Honolulu. This incident has developed a cult following within the EMP science community. The incident allegedly involved blown fuses in a small number (less than 1%) of street light strings. It has been trotted out for decades as the single definitive proof of EMP effects on power grid and long-wire infrastructure."

Hathaway isn't convinced. He presents a painstaking discussion of the incident, subsequent investigations, and the science behind EMP effects. He writes clearly about complicated science, and his conclusion is backed by abundant evidence: "EMP is a ridiculous notion; one that we are supposed to give up our money, our common sense, and our freedom to validate. From the state's perspective, there is always some area of life where people haven't yet developed the proper level of panic to make them tolerate the forced filling of state coffers in relation to that area. There is always something new to fear that the public can't quite grasp without the government to ratchet up its fears."

David Hathaway deserves our gratitude for his excellent and timely account of a little-known propaganda campaign by the State.

Llewellyn H. Rockwell, Jr.
Chairman and Founder of Mises Institute
Auburn, Alabama
December 15, 2017

Introduction

The fear merchants have plenty of foreign interventions to keep them funded at the moment, but they like to have a scary cauldron of witch's brew simmering on the back burner just in case peace breaks out or their lobbyists see a chance to grab additional funds. Enter EMP. Will EMP become the new Y2K? It has many similarities to the Y2K nonsense and a lot of potential for years of taxpayer funded boondoggle technology. This latest fear potion is already being test-marketed in a big way.

Here is the point of this book in one paragraph: Practically everyone who has a public or media platform to speak from believes in EMPs. Even those who are opposed to the government's EMP scare tactics will make their point only while making a disclaimer, saying something like, "Yes, nuclear EMPs are real and cause damage, but we shouldn't be so worried." This almost religious belief is all based on a presumption. This book will show you that the EMP scare is all nonsense. There is no such thing as a high altitude nuclear EMP that will shut down our electrical infrastructure. It has never happened despite thousands of nuclear detonations and it never will.

The single "definitive" event that purportedly proves that the "power grid" and civilian infrastructure can be knocked out from a nuclear EMP is the "Hawaiian Street Light incident." The government funded study that supposedly proves an EMP cause for the Hawaiian incident states in the first paragraph that "Actual (rather than simulated) EMP-interaction data do exist and should be analyzed." After making that statement, the study goes on to make its case using what it properly describes as "imaginary" data. I will thoroughly debunk the conclusions of that study in this book.

EMP harkens back to Y2K in that it is a worrisome sounding phenomenon that is nebulous and hard to nail down. The deeper you look into the theoretical underpinnings for EMP, the farther you will be sent down a rabbit hole pertaining to something like quantum mechanics. At that point, your eyes are supposed to glaze over. You are supposed to give up, throw in the towel, accept it, and move on. The flexible scientific methodology used to justify the existence of high altitude nuclear EMPs is reminiscent of that used in climate science trickery.

The ones who want to gin up fear can only point to a couple of random anecdotes—that are ridiculous when examined—from the heyday of high-altitude nuclear blasts and then speculate about unproven outcomes under modern conditions. So, the story goes, we need to protect against every possibility. We need to "harden" our infrastructure and spend more money to examine how such science fiction weaponry could be used.

Confusing, complex, and unexplainable things work to the state's advantage. Case in point: The Federal Reserve. We are supposed to take the word of the "experts" and allow them to rob us through inflation no matter how shaky the foundation for their conclusions. They are doing things that need doing. Don't try to understand it. You are no expert; and certainly the self-serving experts know what they are talking about.

The same goes with EMP and its effects. Take the word of the EMP scare mongers at Los Alamos, Sandia Labs, Oak Ridge National Laboratory, Lawrence Livermore National Laboratory, and the Department of Energy. But, most importantly, take the word of their crony contractor puppet-masters. Just resign yourself to the fact that other people know a lot more about electromagnetically induced electrical effects than you do.

The scare factory is in full push mode on EMPs. I think it will really get traction when they figure out who can benefit from a cash influx.

Will it be military contractors who can absorb money for the development of protective or offensive devices? Or, public utilities?

Or, will someone propose a big amorphous blob of funding or "EMP tax credits" that public utilities, consumer electronics producers, government contractors, or private citizens can utilize to make or procure EMP "hardened" products? Sorting out the money pie is the key to reaping the harvest from what has already been successfully sown via lots of EMP fear mongering and supporting pseudo-science. Certainly we need an anti-EMP bomb, don't we? Something that shoots out a reverse-phased EMP pulse, right? Science fiction obsessed voters would buy that, right?

That is for Leviathan to figure out. Meanwhile, we can see the hype for what it is and get to the task of proving that this it is a bunch of hokum.

1

EMP Fear Mongering

In case the reader has not heard all the EMP hype, I will hit upon some highlights in this chapter. To kick things off, and to introduce the "definitive" EMP event that is showcased in this book, I will reference an article by Paul Huard on wearethemighty.com that was also featured on businessinsider.com. It was published on November 8, 2016 and was titled, "The first time the US tested an EMP weapon was a doozy." This article talks about the high altitude "Starfish Prime" explosion that occurred in 1962 and the alleged havoc that was rained down upon the Hawaiian Islands and other Pacific Islands from a resultant "EMP:"

> It was 1962 ... people living on the islands dotting the Pacific Ocean from Hawaii to New Zealand were about to see a light show brighter than any July Fourth fireworks display in history.
>
> More ominously, many of those same people would get a taste of how a single nuclear weapon could wipe out a nation's electrical grid – and the U.S. military at the time had no clue how damaging the results would be.

The article describes a scary result: "In Hawaii, the effects were almost immediate: streetlights blew out, circuit breakers tripped, telephone service crashed..."

Huard says that the cause of the failures,

> ...remained secret for years, as did a new discussion that began: how a single nuclear weapon might be used to cripple a nation in one blow.

> It is a discussion that continues to this day as those in the national security community consider how a weapon like Starfish Prime detonated over or near the United States could plunge the country into darkness.

That article says in a nutshell what hundreds of other commentators and state spokesmen say about the current state of the EMP threat and its "Starfish Prime" origins. This book will prove that there was no EMP damage from Starfish Prime and that the whole EMP threat is a massive hoax designed to enrich those who seek control over the federal purse strings.

EMP Advocates Beat the War Drums

The "expert" state mouthpieces and major media outlets are all abuzz about the EMP doomsday machine that allegedly looms large around every corner.

One of them is Dr. Peter Pry who is the Executive Director of the Task Force on National and Homeland Security. He is the official U.S. mouthpiece on EMP issues and is a frequent critic of EMP deniers. He was also a CIA Intelligence Officer from 1985 to 1995.

The Task Force he heads is described as a "congressional advisory board dedicated to achieving protection of the United States from electromagnetic pulse (EMP), cyberwarfare, mass destruction terrorism and other threats to civilian critical infrastructures, on an accelerated basis." Pry is also Director of the United States Nuclear Strategy Forum, which is "an advisory board to Congress on policies to counter weapons of mass destruction." Pry also worked on the "Commission to Assess the Threat to the United States from Electromagnetic Pulse (EMP) Attack" from 2001 to 2008. He is unquestionably the front man assigned the job of concocting fear of EMP on behalf of the U.S. Government.

Pry said in a May 4, 2017 article on Newsmax.com titled "Time to Take N Korea Nuke Threats Seriously" that:

> Almost no one talks about North Korean capabilities to make an electromagnetic pulse (EMP) attack, except to ignorantly dismiss and belittle the possibility...

> ...the mass destruction of electronics and blackout of electric grids over such a vast region would be so injurious that missile reliability matters little — only one nuclear missile needs to work to deliver an EMP attack against an entire nation.

Pry then tries to scare off detractors who oppose his rhetoric and see no need to bang the war drums against North Korea:

> Academics and press pundits, who typically know nothing about EMP, mistakenly assert that a high-yield megaton-class (1,000 kilotons) nuclear weapon is needed for an EMP attack, whereas North Korea's most powerful test was between 20 to 30 kilotons. But a high-yield weapon is not necessary to make an EMP attack.

So, Pry scares us by telling us that a whole nation is threatened by an EMP attack from a single small nuclear device. The actual empirical evidence showing the utter failure of varying size atomic devices—both large and small—to ever produce damaging EMPs, despite thousands of blasts, is discussed later in this book.

He continues:

> A nuclear warhead detonated at 72 kilometers altitude would generate an EMP field with a radius of about 930 kilometers, covering all of North and South Korea and reaching far out to sea. ... This sounds like

the "Gotterdammerung" scenario, discussed by the EMP Commission, where North Korea seeks to defeat an invasion of its territory by an EMP attack that covers most of the theater of operations, to paralyze or cripple U.S. and allied forces, that are far more dependent on advanced electronics and high-tech than North Korea.

Besides the ridiculous claim that there would be EMP damage at this 72 kilometer blast height out to a 930 kilometer radius, which would encompass 2,720,000 square kilometers of territory, the above comment also touches on the unproven theory that advanced electronics are more susceptible to nuclear EMPs than their predecessors. The supposed ever-increasing fragility of modern electronics and ridiculous exaggerations about EMP damage radiuses will be discussed later in this book. The damage radius function is supposedly a function of blast altitude so, similar calculations are often made for even higher altitude blasts resulting in predictions of damage to an area equivalent to the entire lower 48 United States from a single small nuclear device.

Pry continues with a string of EMP scare articles on Newsmax. A more recent Pry article on November 2, 2017 quotes White House Chief of Staff John Kelly saying "That's why he cannot have a ... nuclear device deliverable to the homeland," in response to Laura Ingraham of Fox News making an inquiry about a North Korean EMP attack that would kill "90 percent of Americans."

In a November 28, 2017 article on Newsmax titled "We Can't Afford Missteps in EMP Reportage," Pry says:

> In fact, on March 8, 2005 ... EMP Commissioner Lowell Wood testified before Sen. Jon Kyle's subcommittee on terrorism, technology and homeland security that an EMP attack 'is something which would literally destroy the American nation and might cause the deaths of 90 percent of its people and would set us back a century or more in time as far as our ability to function as a society.'

Pry pounds in the 90 percent American death rate in case you didn't get it the first time:

> Indeed, even before the establishment of the EMP Commission in 2001, during the five years of congressional hearings that preceded the EMP Commission, there was growing awareness and concern that a natural or nuclear EMP event could kill millions of Americans, perhaps 90 percent of the population.

These kinds of statements are clearly made to scare Americans into coughing up the appropriate amount of money when asked to do so for "defense." These "estimates" ignore the fact that there has never been an EMP that has damaged whole systems over wide areas despite decades of blasts detonated at a wide range of altitudes.

On October 13, 2017, Pry used the image of Puerto Rico's hurricane damage to put fear into the hearts of Americans and to market EMP-proof infrastructure in a Newsmax article titled "Tragedy of Puerto Rico Disaster Snapshot of EMP Attack." Pry:

> The tragedy unfolding in Puerto Rico ... is an object lesson in what can happen if the electric grid and other life-sustaining critical infrastructures are suddenly destroyed by a natural or nuclear electromagnetic pulse (EMP), by cyber-attack, by physical sabotage, or by a combination of these. ... No food. No water. Communications and transportation infrastructures so severely crippled that it is difficult or impossible to bring help to those who need it...

In case his audience doesn't feel the tug on their heartstrings, Pry works to make it real on an emotional level:

> Puerto Ricans are rightly afraid for their lives. Mass starvation is not happening, or a pandemic has not swept the island — yet. ... Imagine if Puerto Rico's present catastrophe continued for one year, two

years, or 10 years. That is what an EMP attack would be like.

EMP Advocates Tolerate No Dissent

And since Pry continually blasts EMP deniers, he takes on an EMP critic in the same article:

> Popular Mechanics should send their staff and editorial board to help out in Puerto Rico. Maybe they would stop publishing articles by those who know nothing about EMP mocking Ambassador R. James Woolsey, former Director of the Central Intelligence Agency (CIA), for warning that an EMP attack that blacks-out the U.S. for a year could kill up to 9 of 10 Americans from starvation and societal collapse.

Bring On the Cronies!

And since the EMP fear game is a collaborative effort, Pry throws a virtual high-five to cronies who are working to get proposed state EMP funding into the right hands:

> ... Former U.S. House Speaker Newt Gingrich recently wrote an excellent and visionary article looking beyond the rescue of Puerto Rico to its reconstruction as a showplace, a model for the future of America. One of Speaker Gingrich's ideas is to rebuild Puerto Rico's electric grid so that it is not only modernized — but protected against EMP.

Pry then directly names the one he proposes should manage the crony pipeline of EMP funding while harkening back to a tear-rendering moment of neocon poetic rhetoric:

> ...George H.W. Bush, during the 1988 Republican National Convention, described American voluntarism ... as "a thousand points of light." One of those points of light is my friend Mr. Richard McPherson. Richard is the man who could implement

> Speaker Gingrich's idea of reconstructing Puerto Rico's electric grid so that it is both modernized and protected against EMP and all hazards.

It is ridiculous to somehow make damage from high winds in Puerto Rico into an allegory teaching us that we need to stop denying EMPs and work hard to prevent such EMP-esque damage in the future with lots of taxpayer provided EMP-dedicated bucks. Problems from high winds are somehow solved and remediated by validating EMP? There couldn't be a more extreme apples and oranges comparison. But, sometimes you need disaster and carnage to sell an idea.

It seems as if the spending plan has already been thought out by those in the know. They appear to be merely awaiting the green light from Congress to roll out the tax-funded EMP-hardened model communities of tomorrow. I discuss in this book how this is all a boondoggle cooked up with fake science.

And speaking of Newt Gingrich, he testified on May 4, 2017 to the U.S. Senate Committee on Energy and Natural Resources in a "Hearing to examine the threat posed by electromagnetic pulse and policy options to protect energy infrastructure and to improve capabilities for adequate system restoration."

To establish credibility for his remarks, Gingrich started out by citing the Hawaiian streetlight incident as proof of EMP. That over-hyped non-incident in Hawaii happens to be the main subject of this book. He said, "...I learned that testing hydrogen bombs in the Pacific resulted in burning out lights in Honolulu..."

Gingrich references "a panel of nuclear physicists" called together by Congress in 2004 that concluded that "one EMP weapon detonated over Omaha would cripple half the economy. Further, they found that Russia, China and North Korea were working to develop EMP weapons – and the United States was quite vulnerable to an EMP attack."

Gingrich continues saying that, "...there is no question that the United States should take action to develop a hardened, more resilient electrical system that could better withstand an EMP attack. Frankly, it is a matter of national survival."

He says, "Consider the consequences of hospitals and public safety agencies being without power, communication, or transportation for a significant amount of time." As you can see, Gingrich and other government-connected persons who can attract an audience are working hard to play up the fear angle.

On October 7, 1999, physicist Dr. Lowell Wood testified at a hearing on "Electromagnetic Pulse Threats to U.S. Military and Civilian Infrastructure" in front of the Armed Service Committee of the U.S. House of Representatives. Dr. Wood represented the Lawrence Livermore National Laboratory. Dr. Wood led off with a grossly exaggerated description of the alleged EMP effect from the Starfish Prime blast over the Pacific. This alleged EMP occurrence has become the touchstone event for all EMP advocates throughout the years:

Wood said that this first "EMP,"

> Very unexpectedly and abruptly turned off the lights over a few million square miles of the mid-Pacific. This EMP also shut down radio stations and street-lighting systems, turned off cars, burned out telephone systems and wreaked other technical mischief throughout the Hawaiian Islands nearly 1,000 miles distant from ground zero.

Besides the extreme exaggeration of the geographic scope of the street-light incident (it is only alleged to have happened on Oahu), some of the other things he listed were never even alleged to have happened. From that event, Wood extrapolated that,

> The potential for even a single high-altitude nuclear explosion of a more deliberate nature to impose continental-scale devastation of much of the equipment of modern civilization and of modern warfare soon became clear. EMP became a

technological substrate of the black humor, "Suppose they gave a war and nobody came."

So Wood is suggesting that a single blast would not only end modern civilization in the U.S, but that it would eliminate any possibility for a war response from the U.S. since all modern civilian and military infrastructure would be shut down. Hence the aggrandized statement, "Suppose they gave a war and nobody came."

Wood then paints the picture of a primitive post-EMP America that would result according to official studies:

> When essentially nothing electrical or electronic could be relied upon to work even in rural areas far from nuclear blasts, it was surpassingly difficult to bootstrap national recovery, and post attack America in these studies remained stuck in the very earliest 20th century until electrical equipment and electronic components began to trickle into a Jeffersonian America from abroad.

Wood also said that EMP is,

> ...a continental-scale time machine ... We essentially pick up the continent and move it back in time by about one century ... and you live like our grandfather and great-grandfathers and so on did in the 1890s, until you rebuild. You do without telephones, you do without television, and you do without electric power.

So, per Wood, an EMP would return American lifestyle to that of the "Jeffersonian" era. These types of ridiculous predictions of dire outcomes dominate the landscape of EMP discussions. Even those who poke fun at the ridiculous nature of EMP doom-and-gloom predictions still accept the idea of EMP, presuming that it exists because the court intellectuals have declared it to be so. It is about time that critics examine this presumption rather than merely repeat the state mantra on EMP like good little school children.

EMP Hoax

The above quotes give a small glimpse into the huge volume of EMP crisis commentaries floating around these days from those "speaking with authority" who seek to ramp up the fear of damage to civilian infrastructure. Fear of EMP within the population is necessary to ensure acceptance of any new funding programs that emerge on the horizon to slay the mythical dragon.

2

Even EMP Critics
Validate the Notion
of Damaging EMP

Sharon Weinberger wrote a piece on February 17, 2010 for foreignpolicy.com titled, "The Boogeyman Bomb: How afraid should we be of electromagnetic pulse weapons?" The article properly criticized the ridiculous exaggerations about damage coming from a high altitude nuclear EMP.

Weinberger said that, "EMP has emerged as the latest fear factor-type threat among Washington's doomsday crowd." Weinberger references Lowell Wood's description of EMP as "a continental-scale time machine" and says that,

> ...by talking about 'time machines' and turning the EMP bomb into something that goes bump in the night, those advocating for better defenses risk pushing the issue further into the margins of science fiction.

But, Weinberger is also careful to validate the existence of nuclear EMP, as are other EMP critics, possibly to avoid being described as a kook. Weinberger:

> First observed in a 1962 high-altitude nuclear weapons test over the South Pacific [obligatory touchstone reference to the Hawaiian incident and

31

Starfish Prime blast], EMP is an intense burst of electromagnetic radiation resulting from a large explosion that can potentially wipe out all unprotected electronics. ... The real debate is not so much over whether EMP is a real phenomenon — even critics of the commission's findings agree it exists — but how much of a threat it poses to the nation's infrastructure, how likely it is that a group or country might build and use such a weapon, and what should be done about it. [Bracketed comment added.]

Jeffrey Lewis, an honorable EMP critic that is despised and derided by name by such EMP advocates as Peter Pry has made excellent commentaries and done great research on the EMP "threat inflation industry" as he calls it. He wrote a piece on ForeignPolicy.com titled, "The EMPire Strikes Back" on May 23, 2013 that described the sensational doom-and-gloom falsely attributed to EMP. However, referencing that article in a blog post titled "More EMP Nonsense" on June 10, 2013, even Lewis concedes that EMPs are real, but overblown:

> Although the physical phenomenon of electromagnetic pulse is real, I argued, the severity of an attack is often presented in nearly apocalyptic terms that are simply not supported by the available data.

Nuclear physicist Gordon Prather has also been an outstanding critic of U.S. aggressive militarism in response to non-threats. Prather wrote a piece titled "Bomb-Bomb Iran: To Avert EMP Attack?" on antiwar.com on October 04, 2008. Although Prather makes the point that Iran does not currently have the capability to detonate a nuclear device over the U.S. that could knock out the electrical infrastructure of the U.S., others may.

Prather rightfully criticizes Frank Gaffney for comments made during an interview with Newsmax TV by saying that,

> Gaffney essentially warned that 'any day' now, Iran could detonate exo-atmospherically, somewhere over

Kansas, a specially designed multi-megaton thermonuclear weapon, which could wipe out our entire electricity grid, causing a 'catastrophic disaster.'

Prather says that Gaffney has gone "plain crazy" by saying, in essence, that,

Such an attack [from Iran] could really cripple our 21st-century society, and I would suggest sort of push us back into preindustrial society in the blink of an eye. It would translate over time – not immediately but over time – into the deaths of perhaps as many as nine out of 10 Americans, because our society simply can't be sustained without electricity and all of the infrastructure that supports our urban settings. [Bracketed comment added.]

Yet, Prather, whom I respect tremendously, still concludes in this same article that nuclear EMPs are real and that they are worrisome. His criticism of the Iran EMP rhetoric is based on his assessment that Iran does not have the capability to build the nuclear device and the launch vehicle to deliver it, not that the threat itself—from Russia or China—is not real. Prather:

You see, back in Operation Dominic [the program that encompassed Starfish prime], a series of nuke tests we conducted over the Pacific in 1962, we learned – much to our surprise – then when a multi-megaton-yield anti-ballistic-missile nuke warhead is detonated at the very high altitudes where incoming Soviet nuke warheads would be intercepted, in addition to destroying the incoming Soviet warhead, our ABM nuke's enhanced radiation also produces extreme charge separation in the underlying atmosphere. That is, the atoms in the air are not merely ionized – separated into positively-charged ions and negatively-charged electrons – but gadzillions of those ionization electrons are driven far away from the ions, creating humongous high-frequency dipole radio transmitters. [Bracketed comment added.]

Prather then validates the alleged damaging EMP coming from such blasts:

> The resulting multi-frequency electromagnetic pulse – EMP – can interfere catastrophically with the operation of certain kinds of electrical and electronic systems at considerable distances. That first high-altitude megaton-yield nuke test [Starfish Prime] over Johnson Island resulted in power system failures in Hawaii, more than 700 miles away. [Bracketed comment added.]

So, Prather himself has given a head nod to the "definitive" never-questioned EMP effect in Hawaii based on what was reported at the time as an effect from the Starfish Prime blast. The main purpose of this book is to disprove the claims of EMP effects in Hawaii which have been referenced for decades as the solitary unassailable proof of nuclear EMPs. The alleged "Hawaiian Incident," supposedly involving widespread damage to electrical infrastructure, has never been questioned, even by renowned EMP critics.

That EMP effect in Hawaii is the solitary event used by everyone, EMP critics and advocates alike, to acknowledge the existence of an EMP threat to civilian infrastructure. They use that incident as the touchstone event because it was the only one ever validated by an official U.S. Government funded study. That very flawed study will be dissected later in this book.

The critics in essence say that the EMP threat is there, but that a so-called "rogue state" like Iran or North Korea could not carry out the threat as the official narrative claims.

Prather then validates the conclusions of the 2004, "Commission to Assess the Threat to the United States from Electromagnetic Pulse (EMP) Attack" with regard to Russian and Chinese ability to launch an EMP attack. He says that once the nuclear EMP effect was discovered, the U.S. worked "to see if specially designed nukes of much lower yield could produce EMP as the primary 'kill mechanism.'"

Prather continues:

> According to the Commission, China and Russia have done the same. In fact, in May 1999, during the NATO bombing of the former Yugoslavia, high-ranking members of the Russian Duma, meeting with a U.S. congressional delegation to discuss the Balkans conflict, reportedly raised the specter of a Russian EMP attack that would paralyze the United States.
>
> The Commission concluded that such an attack – non-lethal, in and of, itself – "has the potential to hold our society at risk and might result in defeat of our military forces."

Prather faults the commission for letting the cat out of the bag about the actual Russian and Chinese EMP threat (as opposed to the "rogue state" non-threat):

> Now the Commission has burped again, this time detailing the grizzly details of what might happen if Russia or China or someone with that EMP-capability did attack us. Whereupon, Gaffney, on behalf of frustrated Likudniks, goes on Newsmax to warn that "any day" now, Iran could detonate exo-atmospherically, somewhere over Kansas, a specially designed multi-megaton thermonuclear weapon, which could wipe out our entire electricity grid, causing a "catastrophic disaster."

Prather reiterates:

> Of course, it is one thing for Russia or China to have that capability. It is quite another for a "potentially hostile state or non-state actor" to develop or acquire such a capability.

Prather denounces Iran's ability to do an EMP attack on the U.S. while preserving the concept for the Russians and Chinese:

> Is it conceivable that Iran could develop such a capability? Iran – one of the very largest producers of oil and natural gas in the world – can't even construct the refineries it needs to produce enough gasoline for its population.
>
> In any case, if Iran was to somehow acquire a multi-megaton nuke (from the Russians or the Chinese?), why do the Likudniks think you're stupid enough to believe their claim that the Iranians would choose some non-lethal use for it? Like using a magic carpet to haul it up 50 miles or so above Kansas and detonating it? Is that what you'd do? Or would you use eight tiny reindeer?

Prather properly decries the suggestion of a "preventative strike" against Iran while not shutting the door on an EMP coming from the Russians or the Chinese:

> Do the Likudniks really believe that you're stupid enough to believe that if the Israelis don't launch a "preventative strike" against the Mullahs and the Iranian IAEA-Safeguarded facilities this year or early next year that nine out of ten of us will – if we're the lucky ones – freeze in the dark?

My purpose here is not to criticize Prather's huge body of fine work, but to merely point out that even those who downplay the EMP threat (like Prather, Lewis, and Weinberger), still believe that it is a real concept that it is looming on the horizon. My point, and the point of this book, is that the whole thing is fake and doesn't fit with real science. And, no, as you shall see, there was not a nuclear EMP in 1962 that fried the electrical infrastructure of the Hawaiian Islands or other Pacific Islands (as most every EMP writer claims) as the result of "Starfish Prime," the mother of all high-altitude blasts.

The scientific legitimacy of the threat is rarely questioned, or even examined. It is about time for that to happen. Someone has needed for a long time to take on the task of reviewing the official research and the scientific claims—on both a scientific and practical level—and see if there is any truth to them.

3

Alleged Failure of
Hawaiian Street Lights

At nine seconds after 11 PM on the night of July 8, 1962 (Hawaiian time—it was July 9 in the continental U.S.), the "Starfish Prime" atomic warhead was detonated over the Pacific. Residents of the Hawaiian Islands stayed up that night and watched night turn to day as the sky went up in flames from the largest outer-space nuclear blast in history. The blast was 93 times more powerful than the one that destroyed Hiroshima. One of the U.S. Government's main stated purposes for the blast was to study EM (electromagnetic) effects on communications, electrical devices, and electrical infrastructure.

A damaging nuclear "EMP" (ElectroMagnetic Pulse) effect had been postulated since the first atomic blast in 1945, usually for long-wire infrastructure. Evolving theories later suggested that high-altitude blasts would exhibit an enhanced EMP effect (sometimes known as HEMP: High Altitude EMP) also affecting long-wire infrastructure. But, no electrical infrastructure damage had been proven from any of the numerous high-altitude nuclear blasts conducted by the U.S. and the U.S.S.R.

Starfish Prime was the chance to determine what the EM effects would be. It was the most thoroughly instrumented nuclear blast in U.S. history. Many special-purpose ships, planes, missiles, and ground stations looked for EM effects. The U.S. Government scoured a vast swath of territory within the radius of predicted EM effects looking for any damage. This area of predicted effects included numerous populated Pacific Islands. The media was clued in and looking also.

The resultant singular and "definitive" EMP event, which allegedly gave documented proof of EMP-caused infrastructure failure, became known as the "Hawaiian Street Light Incident." The alleged incident happened on the island of Oahu which is made up of the City and County of Honolulu. This incident has developed a cult following within the EMP science community. It has been magnified over the years to the point where many claim that lights were burned out on all of the Hawaiian Islands (not just Oahu) and on many other Pacific Islands as well by an EMP. Some even claim that the electric power grid was knocked out in the Hawaiian Islands or in other Pacific Islands, which was never even alleged to have happened.

In actuality, the incident allegedly involved blown fuses in a small number (less than 1%) of street light strings on Oahu. It has been trotted out for decades as the single definitive proof of EMP effects on power grid and long-wire infrastructure.

With a total of 2,476 nuclear blasts around the world, this is it. Nothing like a mass-damage EMP effect has ever been seen. EMP proof—backed by official science—has to lie with this event because it is the "definitive" EMP event and the closest thing to a "proven" mass damage event they have after thousands of blasts.

For the U.S. Government, looking back at the event, it was now or never. The potential for wide area civilian infrastructure damage from EMP had to be present in this Hawaiian event or the theory would die on the vine and be relegated to joke status alongside Bigfoot. This could not be an isolated single failure. It had to be a scary community-wide failure that impacted the whole island of Oahu and the city and county of Honolulu. Societal fear must be generated from this "occurrence."

That event and the official study funded by the U.S. Department of Energy which "proved" a nuclear EMP cause are the primary subjects of this book.

I will show that the pre-ordained EMP conclusion is completely inaccurate because A) Any electrical failures in Hawaii were statistically insignificant and do not indicate a mass-damage effect, B) The study was completely flawed in that it reviewed no data because there was no data to review, C) The shoddy "third world" street lighting circuitry in Honolulu at the time was prone to internal failures that had nothing to do with EMP, and that D) The EMP theory itself related to the Hawaiian event and other situations is theoretically impossible.

This book starts out with an explanation of EMP theory and the flawed physics involved. The theory will become useful to the reader when he delves into the actual Hawaiian incident and other discussions later in the book.

A suggestion for some readers:

The real nuts and bolts of the Hawaiian incident and the massive fraud perpetuated by the official study are contained later in this book. I have tried to present the correct science in the following section in a way that is easy to grasp. On the other hand, the government has purposely made its version complex. Some readers may prefer to not get bogged down in a conversation about the correct physics of electromagnetic induction contrasted with the flawed physics used by crony contractor researchers and government personnel to prove the existence of EMPs. Those foundational ideas are presented early in this book to aid readers in discerning the flaws in the official study of the Hawaiian event. Readers who would prefer lighter reading, but want the gist of this book, may choose to skip the discussion of theory and go straight to the Hawaiian street light incident and the flawed official study. If so, you may go directly to Chapter 11.

4

Electromagnetic (EM) Induction

How Do Electromagnetic Waves Induce an Electrical Flow?

To fully explore the fallacy that a nuclear EMP can induce a damaging electrical force at a long distance in a circuit or in a conductor like a wire, we need to understand how EM waves can actually induce electrical forces in the real world. It does happen. Only by understanding this, can we see how government-affiliated researchers have distorted the science to fit the outcome they desire.

Directional electrical flow can be induced electromagnetically in conductors that are not physically connected to the source of the EM waves that is exciting the flow. The directional force and resultant directional flow is key.

Photon Bombardment Does Not Produce Directional Electrical Pressure

A large amount of energy from the sky and from other sources in the form of photons and free electrons is continually impacting things around us including metallic objects like power lines, phone lines, street light wiring, metal roofs, etc. Wave-configured photons arrive in the form of sunlight, gamma rays, etc. Free electrons and scattering photons result from photon collisions with atoms in the atmosphere and on the Earth's surface. But, they are not exerting a directional pressure and propelling a directional flow of electricity down those wires or through other metal surfaces.

Photons impacting a wire cause electrons to be momentarily ejected from atoms in various directions. The free electrons cancel each other out as they fill other electron "holes" made by other photon impacts and there is no directional flow in the wire. Their effect is inconsequential because it is self-cancelling. You can't just string up a long wire between telephone poles and receive directional electrical energy from the continual bombardment of your wire by photons. Neither can you connect a wire to each end of your metal roof and harvest electricity. The photons arrive continually and eject electrons within your wire, but they don't push electricity through your wire.

How can we make an electrical flow in a conductor? We can restrict the movement of the ejecting electrons so that they can only move in one direction through the conducting material which needs to be arranged in a loop. Electron "holes" being produced in one end of the material will then be filled by free electrons arriving in a loop as the conductive material carries ejected electrons in only one direction to fill the holes. A directional flow in the circuit will result, i.e. a current.

Lack of current directionality from photon bombardment of wires is one of the problems with nuclear EMP theory.

Making Directional Current from Random Photon Bombardment

Directionality in a circuit can be obtained through the use of diodes that block the flow of electrons going in one direction, but not the other. Directionality can be obtained in photovoltaic solar cells by the use of a single-direction crystal lattice that acts as a diode and blocks the electrons that are being randomly ejected by the photons in sunlight from flowing backwards through the lattice. It only allows the free electrons to rebound in one direction from the bombardment of photons. The "photoelectric effect" can be harnessed and turned into an electrical flow when the electrons are only allowed to move in one direction. Other forms of current "rectifiers" allow changing electron flows in a circuit to move in only one direction. Directional flow is not imparted upon wires by the random self-cancelling scatter of photons and electrons occurring after a nuclear blast.

Also, electrostatic effects like lightning that involve direct electrical flows through the atmosphere should be differentiated from electromagnetic "EM" effects we are considering here which utilize waves of photons.

Electrical Force from EM "Flux"

The "flux" issue described below reveals another hole in nuclear EMP theory.

Electromagnetic (EM) waves can induce an electrical potential (voltage) and a corresponding electrical current in good conductors, typically metal. However, in order for EM force to produce a current, a flow, in a conductor—let's say a wire—the EM effect cannot be constant or a single event. It must present itself in a state of flux, i.e. continually switching on and off, continually changing polarity, or continually moving back and forth. For example, a stationary bar magnet cannot induce an electrical flow in a nearby wire.

Neither can the EM force produced by a wire-wound electromagnet induce such a flow; *unless* the EM force is "fluxing" in relation to a nearby aligned secondary circuit. A wire-wound electromagnet can produce a charged electromagnetic field, but the field cannot cause a parasitic directional flow of electrons in another wire unless the source EM field is turned on, and then off, and then on, and then off, etc. causing "waves" of photons to emanate from the electromagnet that propel electrons incrementally through the secondary circuit; or, the EM field can be moved back and forth and back and forth, etc. in relation to another wire; or, a third possibility, the current powering the EM circuit can be continuously reversed back and forth and back and forth (i.e. utilizing alternating current or "AC" in the parent EM circuit) which would reverse the polarity of the EM field in regular cycles. The flux principal is the essence of Faraday's Law of Electromagnetic Induction. Without it, no electrical flow can be produced in a wire from EM energy.

Electrical Flow Induced by EM Flux

Small wire-wound transformers in the wall chargers for cell phones or larger transformers inside larger, typically older, electronic devices like clock radios and TVs induce a desired voltage and current flow (an electromotive force) in a secondary circuit not through direct electrical contact, but by EM waves that are continually changing; i.e., in a state of flux. The flux is obtained via the AC power coming from your home electrical outlets that is changing the polarity of the transformer's EM circuit 60 times a second.

A similar current-inducing EM effect can be obtained via a DC "chopper" circuit that turns an EM circuit on and off rapidly or via a mechanical shifting of positions of a man-made electromagnetic field (usually from a movable EM coil) that is changing position back-and-forth or circularly in relation to another parasitic coil that then becomes the recipient of an electromotive force from EM waves, not from a direct physical connection. The shifting EM field could come from a rotating wire-wound metal core in a motorized generator that is continually changing its presentation angle to another nearby (but physically unconnected) secondary circuit.

But, the important conclusion from the above is that EM electrical induction happens from a continual unified shifting of a group of EM particles changing in unison causing their effect to be imparted on the secondary circuit in separate *waves* of photons. All pushing one way, then the other way, etc. Or, on / off / on / off, etc. This EM force shifts the electrons in one direction and the flux then causes the circuit to reset and push more electrons in a stream behind the first electrons in ratcheting incremental fashion down the wire.

Electrical Flow in Wires That Are Not Connected Physically to EM Source

All of these states of "flux" in a polarity-aligned EM field could cause a parasitic flow of electrons in a nearby wire that is not connected physically (only electromagnetically) to the parent EM circuit. Something equivalent to a diode could then block the shifting electrons from moving backwards in the circuit when the EM force is fluxed to ratchet another batch of electrons down the wire. The diode acts like a one-way valve. That is how voltage transformers and radio devices work. Alternating current changes (fluxes) the magnetic field many times a second, propelling photons in waves across a gap that push electrons incrementally in waves down a wire in a physically separate parasitic circuit.

Radio waves can only propagate and induce a feeble current in a distant wire (antenna) because of EM flux. Without the flux, they would not propagate. Radio waves are electromagnetic (EM) waves that are generated in an alternating polarity-switching back-and-forth AC fashion. They are coherent waves of photons; not random scattering particles. When a radio signal induces a current in a distant metal antenna connected to a sensitive receiver, it only occurs because of the flux that creates the coherent waves. There would be nothing to propel an electrical current through a distant conductor (wire) separated by an air gap without the flux which creates a repeating series of EM waves moving through the atmosphere which push the electrons incrementally through the conductors in the receiving radio's antenna and circuitry. Radio receivers also have diodes / rectifiers to make the alternating current directional in the receiver circuitry.

This current-inducing EM flux has not been demonstrated in a nuclear EMP. A directional electrical force cannot be induced in wires from photons arriving from high-altitude (after a nuclear burst) without EM flux or something to inhibit backwards flow in the wire from random photon bombardment.

Photons Arriving from Nuclear Blast are Random, Not Coherently Fluxed

The lack of EM flux and the lack of directionality are two strikes against nuclear EMP theory.

Particles arriving on Earth from collision and ejection of electrons and photons in the atmosphere from a nuclear event do not arrive in coherent wave fashion. Impacting photons are acting as particles at uncoordinated intervals; i.e. out of phase with each other. That is why, once again, a long wire on a power pole doesn't receive a current from the constant bombardment of photons that is impacting it every day. There are a lot of photons, but they are uncoordinated. They are not arriving in coordinated "fluxing" waves that can push electrons in a current through a wire. [*Note: The plasma frequency of*

47

the metallic conductor being lower than the frequency of any wave-configured photons is another reason for a lack of induction in a wire. This will be touched on later.]

The way this relates to nuclear EMP theory is that there is only one of each type of supposed EMP pulse: on and then off. Even that pulse is not a coherent coordinated wave. If it exists, it is made up of particles emitted from the blast and others that later impact Earth at random after multiple atmospheric collisions. There is no flux to push a current through a conductor as there is in a typical pair of EM inductive circuits where a primary circuit uses regular EM pulses emanating from an electromagnetic coil to induce an electron flow in a secondary circuit that is also made of a long wire wound in a coil (as in a transformer).

The Hawaiian street light incident was allegedly caused by a current flow in wires from an EM force, but neither the EM "flux" needed for EM induction nor a directional photoelectric effect have been demonstrated to explain the supposed nuclear blast origins of any alleged current flow in the wires.

5

EM Emissions Versus EMP

Electromagnetic, electrical, and random photon / electron emissions exist in nature and can also be man-made. Normal solar radiation and solar flares produce amounts of scattering / colliding electrons and photons in the atmosphere well beyond anything man can produce.

EMP theoreticians admit this, but then embark on a debate about the duration of alleged single nuclear photon "pulses" and add the EM (electromagnetic) designator to them all, even if unwarranted. "EM" to some extent is a misnomer when talking about any electrical effects from photon emissions occurring during a nuclear blast, since "EM" implies on / off "fluxing" of groups of photons acting in coherent waves at specific frequencies. The photons from a nuclear blast are usually described initially as gamma rays and then as scattering particles after they have impacted atoms in the atmosphere and have ejected electrons. Even if a single "pulse" was "electromagnetic" and had the correct changing orientation to all portions of the recipient looping circuit (impossible since the blast is at a certain azimuth to the wires), it could only introduce an electrical potential in a wire as a single pulse and not an electrical flow that would blow fuses as in the Hawaiian incident.

A Pulse Is Not a Wave

The "P" in EMP stands for "Pulse," which is misleading. The word "pulse" implies "wave," which suggests the multiple fluxing EM "waves" of photons needed for a functional theory in keeping with

Faraday's Law. Some guilt ridden EMP-advocating scientists have apparently felt the need to describe the emissions in a way that could be portrayed—if the need ever arises—as arriving in "fluxing" EM waves, making them able to excite parasitic currents in wires.

Otherwise, they are just random scattering photons that could not induce a directional flow of electrons in a piece of metal (like a long wire electrical transmission line). The flux issue is mainly unstated (and crisis advocates certainly don't care), but it can be inferred from other scientific hijinks aimed at getting an EM frequency for the "pulse" (discussed later). So, they keep the "P," which suggests waves, and try to use Faraday's science of EM flux to prove the potential for damaging EMP effects to make the "threat" believable.

E1, E2, and E3

They distinguish between three categories of alleged nuclear EMPs that would supposedly occur during each blast: E1, E2, and E3. Each one is allegedly a single pulse and not a fluxing event. E1, they say, is the shortest pulse. The pulse length of E2 is allegedly analogous to the length of the electrical effects from a lightning strike, and E3 an allegedly even longer pulse similar to a solar flare. Per conventional EMP theory, long wire systems (e.g. power lines, phone lines, street light wiring) are described as those being susceptible to EMPs in that they would supposedly receive an induced electrical flow per the theory of electromagnetic induction. E1, E2, and E3 have all been associated in EMP scientific literature with theoretical voltage / current excesses in long-wire systems, although E2 and E3 are described more often as being the culprits in the long wire context. E1 is mentioned also, speculatively, as a phenomenon that will fry modern electronics.

6

High Altitude Nuclear EMP: Basic Theory and Problems

A high-altitude nuclear EMP (also known as HEMP) is supposedly produced when a high-altitude nuclear burst (like the high-altitude 1.4 megaton Starfish Prime explosion in 1962—examined later) produces photons in the form of gamma rays that strip electrons off gases in the atmosphere. Scattering photons and free electrons then allegedly push their way into other nearby atoms causing successive scattering. Aspects of this process are described in the theory of "Compton Scattering."

Sub-Atomic Scatter

It is hard for the free electrons to find purchase as an "electrical wave" in the atmosphere (which is a very poor conductor) or on the distant ground, so they die out / are re-absorbed in the atmosphere quickly as they move towards equilibrium by bonding with atoms. But, the scattering particles and photons, allegedly, after a nuclear burst, produce an electromagnetic "pulse" on a different plane to the electrical wave that can induce parasitic electrical currents in a correctly oriented distant conductor, like a phone line or a power line. That is one common generic explanation; with sort of a built-in assumption that everything would just pile up in a recipient wire and create a big directional current flow; even though there is no flux to cause a flow.

Particle-Wave Duality of Photons

Even modern quantum physicists are not in total agreement when it comes to the particle-wave duality of photons. Photons may be perceived as particles or waves under different circumstances. They can be scattering particles or moving in bunches, in waves.

It is one of the defining issues of quantum mechanics. It is also the key to debunking popular nuclear EMP theory. Scattering photons after a nuclear blast are not moving in groups, in coherent waves, that can induce directional voltage and corresponding current flow in a parasitic circuit; they are acting like particles. Coherently fluxed / pulsed EM waves (as in transformers and RF circuits) can exert an electric potential, a voltage, and push a current incrementally through a circuit.

This, in my mind, is the basic problem with EMP theory. Any photons arriving from scatter (photon emissions from impacted atoms or particles in the atmosphere) are not arriving in coherently organized fluxing waves that can propel electrons in stages through a wire. And any photons grouped into gamma waves cannot push a current through a wire because of the plasma frequency issue discussed next.

Although gamma rays are described as photons having a wavelength and a corresponding frequency, they are usually described for HEMP purposes in their function as photon particles acting individually to impact and strip electrons off atoms in the atmosphere. Unless the photons are part of a series of EM pulses acting in coherent waves, they cannot induce an electromotive force (in a wire, etc.).

Lower Plasma Frequency of Metal Prevents Induction from Higher Frequency EM Source

Photons that are said to be moving in waves (and hence described as part of the EM spectrum) do not induce currents in metallic conductors when the "frequency" of the EM waves exceeds the "plasma frequency" of the metal.

The photons of concern emanating from a nuclear blast (for HEMP purposes) are described as being in the form of gamma waves.

Gamma waves have a frequency in excess of 10 exahertz, giving them a wavelength smaller than an atom. Because the frequency is higher than the plasma frequency for metals used as conductors in wires, a current cannot be induced in wires from these EM photons grouped in waves at the high frequency of gamma waves.

Therefore, gamma rays (although considered to be EM waves) that bombard long wire systems do not induce an electrical flow. Because of this, theoretical gymnastics are occasionally introduced by contract EMP scientists to obtain lower EM frequencies for photon emissions that are low enough to induce electrical flows in those pesky long-wire systems on Earth. [I can imagine them laughing all the way to the bank.]

The desired lower EM frequency photons are postulated theoretically as a later effect after the initial gamma bombardment of atoms in the atmosphere. This explanation is deployed if you dig deep enough into speculative EMP theory. It is discussed later in the section titled "Magnetobremsstrahlung."

7

"Deep" HEMP Theory

Particle Wavelength Fiction

The need for coherent EM waves for the induction of electromotive force in secondary conductors has caused extraneous "wavelength" fiction to develop in regard to high altitude nuclear EMPs (HEMPs). As has occurred with climate science, nuclear EMP science has been plagued with evolving nonsense to fit the desired outcome (complete with a lot of complex formulas to suggest legitimacy and to scare off detractors). So, how have they fictionalized the emissions from nuclear bursts to fit with EM induced currents in human electrical systems?

They have begun to talk a lot about "wavelengths" of sub-atomic particles and photons to get their desired EMP effect. This is a bogus analysis designed to create a presumed relationship with "frequency" which is usually the bedfellow of wavelength in physics and in electronic theory. Frequency is the rate at which distinct on/off waves of photons are emitted from the primary EM source that will induce a current. The problem is that there is no frequency for scattering particles because there is no EM wave flux cycle that will induce a directional electrical flow in a parasitic circuit.

The talk of wavelengths for photons entering into and emanating from atomic collisions morphs automatically into talk of associated frequencies—even though the frequencies (if described as such) are actually uncoordinated, chaotic, and particle specific. A wave frequency is needed, though, to make a calculation of induced electromotive force in a recipient circuit per Faraday's Law. An unwarranted, inaccurate, and inappropriate reverse engineering of an EM frequency is sometimes obtained from what are not actually wavelengths between regular emissions of coherently grouped particles.

EM Wavelength: Radio Analogy

You can calculate an EM wavelength from a frequency or an EM frequency from a wavelength if you have regular on /off pulses and a constant speed of travel. The wavelength is the distance between batches of photons that are travelling in multiple separate waves at the speed of light. For example, a radio frequency of 1,000,000 cycles per second (1 MHz) would have a wavelength of 300 meters between each wave. After each EM radio wave (coherent group of photons) exits the transmitter antenna, it travels 300 meters before the next wave exits the antenna.

Bullet Analogy Regarding Wavelength

Another example of the relationship between wavelength and frequency can be seen when examining the distance through the air between bullets from a machine gun firing at a specific cyclic rate (frequency). The wavelength is the distance between the bullets. The "wavelength" between 7.62mm bullets coming out of an M-14 at 750 rounds per minute is longer than the wavelength between those same 7.62 bullets coming out of an M-134 Mini-gun at 6000 rounds per minute. The distance in the air between the bullets emanating from the Mini-gun represents a shorter wavelength due to the higher firing "frequency." The machine gun analogy is imperfect when discussing EM wave effects though, because each bullet is an individual item and not part of a group of bullets travelling together in separate waves.

Misuse of Particle Wavelength Leads to Presumption of Coherent Waves of Photons

How is particle wavelength misused to advance nuclear EMP theory? After concluding that non-specific photons have a specific "wavelength" and thus a reverse-engineered specific frequency, an unexplained leap in logic is made when it is presumed that all of these uncoordinated photons automatically constitute coordinated EM waves with matching polarities moving in unison. They could then induce an electromotive force across an air gap in wires on Earth. This connection is presumed merely because they have assigned a frequency to what amounts to individual particles.

In actual effect, these scattering photons are uncoordinated, incoherent, and rebounding in random directions. For electrical induction purposes, they cancel each other out because they are random. It is hard to conceive of the frequency of individual photons that are not acting in waves. Such random particles do not induce electrical currents in long wire systems. Random atmospheric photons may interact with long wire systems all the time, but they do not induce a directional electrical flow.

Magnetobremsstrahlung

What has been described so far is the basic theory, but some scientists, apparently seeing a hole in the theory similar to what I have noted, go a little further down the trail of quantum mechanics and proffer an argument for a nuclear blast effect they say *would* produce coherent EM waves from the scattering photons and ejected electrons at a frequency low enough to induce a current in a wire. The theory of magnetobremsstrahlung (literally: magnetic braking radiation) or cyclotron / synchrotron radiation is used to postulate the final part of a chain of presumed events that would allegedly produce useable EM waves—useable from the perspective of inducing electrical flow in wires on Earth from an HEMP. The following is what they allege.

First, the photons emanating from the blast would collide with atoms in the atmosphere ejecting free electrons. The electrons would then descend towards the Earth moving along the lines of the Earth's magnetosphere and emit radial EM radiation (different photons) as they begin to spiral due to magnetic braking. These are not the same photons at the gamma frequency that emanated from the blast.

This, they suggest, would create the EM flux (coherent on/off groupings of photons) to fit with Faraday's Law of EM induction at a low enough frequency to preclude an objection regarding the plasma frequency of the metal in the wires. But, the theory of magnetobremsstrahlung describes how individual electrons could create EM radiation. It doesn't necessarily explain how *all* electrons flowing towards Earth would create EM radiation in *unified waves* that all radiate and arrive *simultaneously*; how the EM frequency allegedly emanating from each now-spiraling electron would match and coherently work in conjunction with the EM frequency generated by all the other spiraling electrons ejected from atoms by photons coming from the nuclear burst.

They would all have to emit radial radiation exactly in phase with each other, or they would be self-cancelling and not pushing in the same direction in successive coherent waves, i.e. fluxing. This is the only way they could push electrons through a wire on Earth. To use a machine gun example again: A thousand machine guns firing at a shooting range would be out of sync. The bullets would not all be leaving the barrels of the different guns in unison. Also, they would not all necessarily be firing at the exact same cyclic rate; the same frequency. There would not be coherent waves of bullets flying through the air with a clear wavelength between each wave.

So, they pretend to have found a way to bridge the theoretical gap from photons acting as scattering particles (or at gamma frequencies that are too high to excite a current in a wire) to photons acting in lower frequency EM waves. All this is accomplished by saying, in essence, that the initial photon collisions with atoms in the atmosphere produce ejected electrons that then move along Earth's magnetic lines, which causes them to spiral and produce radial EM cyclotron radiation, i.e. different photons in waves. So, they go from gamma wave configured photons to scattering photons, to electrons, and then back to photons, but allegedly now in the form of coherent EM waves at a low enough frequency to induce a current in a wire on Earth.

An alternate version of this high altitude EMP theory abbreviates the explanation by saying that initial Compton Scattering from the blast results in radial radiation from positive / negative "charge separation" occurring in a dipole pattern along Earth's magnetic lines resulting after the blast.

Don't Try to Understand

Now would be a good time for an interlude to recall what I said in the beginning about the Federal Reserve, climate science, and other such "complex" things that we are supposed to take for granted. They are just too difficult for us mere mortals to grasp. The state just needs to "handle" those things on our behalf. That is really the point of the theory section of this book: To get the reader to see the mental gymnastics that will be deployed by court intellectuals to get him to accept the ridiculous EMP presumptions as valid—because they are beyond his ability to comprehend.

They don't comprehend them either. They are not unified. Why? Because the theories don't work despite all the tweaks. Even if a crony physicist argued with me on some of the points, the discussion would be meaningless because there are so many iterations of these nonsensical theories. He could just add a delta symbol and a couple more variables to an equation and say, "See, it works now." What actually matters is the pudding. The whiteboard theories can always be modified and haven't produced anything. The proof is in the pudding. This nonsense does not make pudding. An "EMP" has never caused a mass damage effect over a wide area despite thousands of nuclear detonations, and it never will. That is because the theories have no bearing on reality.

8

Problems with Magnetobremsstrahlung / Cyclotron Predictions

Magnetic Conjugate Regions

After embracing magnetobremsstrahlung and resultant cyclotron radiation as the silver bullets to get lower frequency EM waves out of the scattering particles produced by a nuclear blast, they then conclude that nuclear EMP effects on long wires, etc. would not only be massive area-wide events spanning entire continents, but that they would be felt the most in "magnetic conjugate regions" on Earth or in areas with stronger magnetic fields.

Why No Free Cyclotron Power from Solar and Cosmic EM Radiation?

If this effect worked to produce coherent EM waves at a low enough frequency to induce directional electrical flow in long wires and metallic conductors on Earth, we would see it happening already in those same wires from normal solar and gamma radiation that also causes electrons to be ejected from atoms in the atmosphere. Why don't those electrons then spin along the magnetosphere making radial radiation and produce a directional flow in the power lines, phone lines, and street light wires on Earth as we speak? We would have directional electrical flows on Earth and even stronger directional flows in wiring at magnetic conjugate regions.

We could then conceivably be harvesting the power generated by all the displaced electrons produced by bombardment from solar and cosmic photons that would be aligning themselves along the magnetosphere. We could do so by merely stringing up long wires or sheets of metal without the need for photovoltaic solar panels. The electrons would be spinning, throwing off coordinated radial EM emissions and making current flow in wires.

Conversely, if you like the charge-separation theory better, the positive and negative ions produced by continual day-to-day gamma bombardment would be separating and aligning their polarities along the Earth's magnetic lines and would be producing radial radiation in a dipole pattern that should also be producing directional electrical flows in wires on Earth. None of this happens.

There is another problem with this ludicrous notion; and that is related to the direction that these theoretical EM waves are sweeping across looping long-wire circuits (discussed later).

No Failures in Magnetic Conjugate Regions

The U.S. and the Soviets actually did research at both northern and southern magnetic conjugate regions during high-altitude nuclear blasts. They hoped to see enhanced EM effects. The Soviets, without asking permission, even snuck two EM research ships into the same Northern and Southern Pacific magnetic conjugate regions being studied by American ships and ground personnel during the Starfish Prime blast to observe the EM effects in these zones that would supposedly receive concentrated effects. No mass damage or large effects were reported by either the human populations living in Pacific magnetic conjugate regions or the research personnel in those areas during or after Starfish Prime. The only significant observations in the magnetic conjugate regions pertained to a distant visible aurora in the sky from the nighttime blast. In short, power was not knocked out, ships weren't disabled, and planes didn't fall out of the skies in those regions that were supposed to receive severely heightened EMP effects.

So, there were no mass failures or notable consequences—even in magnetic conjugate regions—as is postulated by this "deep" EMP theory that we aren't supposed to even try to understand.

If you have read a bit about "EMP," you have no doubt read speculations that magnetic fields would result in a concentrated EMP effect on the continental U.S. and in Europe and Asia. Rest assured, this concentrated effect has never happened, even in regions that were supposed to receive a heightened impact during U.S. and Soviet blasts. Research personnel have been sent in the past to the areas where magnetic lines would supposedly draw in the most concentrated effect from high-altitude tests. There was no mass damage effect.

Overestimation of Magnetobremsstrahlung Effect

Of course, the proper exponentially reduced energy levels are not fully calculated for this theoretical chain of events. EMP advocates attempt to calculate the total EM energy that would theoretically emanate from a nuclear blast of a certain size, and then claim that the EM energy would impact the Earth's surface and couple with human electrical systems with a lot of that energy. There is no full accounting for an exponential reduction of strength from dissipation, even if we take for granted that there would be coherent EM waves produced by such an effect that could induce electrical currents in distant wires.

They fail to calculate how the initial photon energy is cut dramatically at each step. Energy is lost when the gamma photons collide with atmospheric atoms. Then, most of the ejected free electrons (that are not carrying all the initial energy from the blast), almost immediately bond with atoms. Any remaining free electrons that would allegedly align themselves along the magnetosphere would be infinitesimal and carrying a vastly reduced amount of energy from that calculated for the initial blast. If those electrons were to spiral and produce radial cyclotron radiation, the new EM radiation of photons at a now lower frequency would be at a tiny fraction of the energy level of the surviving ejected electrons that flowed along the magnetic lines. That is, the spiraling electrons don't disappear and pass on all their energy to peripheral photons that emanate from the spiraling process. The electrons still have energy and most of them ultimately bond with atoms giving up their free electron status.

The only energy given up by the electrons to photons comes from a reduction in their speed from magnetic braking that causes them to spin rather than only moving forward. Recall that magnetobremsstrahlung means "magnetic braking radiation." So, the only imparted EM energy is from the change in motion and speed as the electrons change to more of a spiraling motion from a straighter motion. The electrons haven't given up all their energy. The electrons still have energy. The Earth's magnetic field only causes them to change their movement (allegedly) which would make them give off an exponentially smaller amount of photon energy compared to the actual energy they retain. Even this new much lower energy cyclotron radiation (if it exists after the blast) is not radiating in one direction only. So, it is dissipating exponentially if energy density is considered for an expanding sphere of EM waves.

Also, as touched on before, any vastly weaker cyclotron photons (weaker when compared to the total energy of the initial gamma photons) would not necessarily emanate in unison at a specific frequency coinciding exactly with that of all the other photons emanating from any other ejected electrons spiraling along the Earth's magnetic lines. In that case, they would not produce coherent EM waves and any effect would be even weaker or self-cancelling.

9

Problems with Intensity and Directionality

Reductions in Terrestrial Nuclear EM Impact from Radiation Belt Left in Space

There is also not a sufficient reduction calculated for the significant amount of energy left in space in the form of artificial radiation belts that remain for years after high-altitude nuclear blasts. These long-lasting radiation belts are an acknowledged effect and were studied by U.S. and Soviet satellites after each blast. Much, or most, of the energy from outer space blasts (like Starfish Prime) goes in the direction of least resistance—up or sideways. The energy dissipated into space and left behind in radiation belts should be subtracted appropriately from the total energy that would allegedly impact Earth in the form of an "EMP." This deduction for energy going upward and outward is essential if we are doing honest science, since an EMP effect is only postulated from high-altitude blasts where this would always be a major issue.

Reduction in Nuclear EM Impact from Expanding Blast Sphere

And finally, in the category of exponential reductions in energy, the energy remaining in each portion of the expanding sphere from the blast gets exponentially smaller as the diameter increases. As the diameter of the sphere gets large enough to span the entire continental U.S., the surface area of the expanding sphere becomes enormous. Energy density plummets along with this larger sphere.

Any remaining energy that could supposedly fry power lines and electrical infrastructure in each part of that expanding sphere has, in actuality, become infinitesimal.

If calculations are made of the EM energy per square foot, the differences become huge. A sphere of 200 miles diameter around the blast would have an approximate surface area of 125,700 square miles. A growing blast sphere that now has a continent-spanning diameter of 2,000 miles would have an approximate surface area of 12,570,000 square miles. A ten-fold increase in the diameter of the blast sphere results in a hundred-fold increase in the surface area— and an exponential reduction in energy conveyed by each portion of that vastly larger surface.

It is ludicrous to conclude, as is done frequently, that a single nuclear blast over the central U.S. would have enough retained EM energy density to induce massive current in power lines and to fry electronic devices in Los Angeles and New York after its blast sphere has expanded all the way to the West Coast and East Coast. Each portion of the expanding sphere would lose energy density exponentially as the surface area increases from hundreds of square miles, to thousands of square miles, to millions of square miles, and then to tens of millions of square miles.

Problem: Wire Loop Cancels Directional EM Waves from Outside of Loop

Yet another problem exists with nuclear EMP theory. I have decided to place this discussion after the discussion of cyclotron radiation / magnetobremsstrahlung because it comes into play after we assume that we have coherent EM waves (we don't) arriving on Earth as the result of a nuclear blast that are at a low enough frequency to induce electrical flow in wires. Cyclotron radiation allegedly gives us those lower frequency EM waves of incoming photons from spiraling electrons, although, if it did, the EM energy would likely be random, uncoordinated, and infinitesimal.

For the sake of discussion, I will assume that we have now come up with EM waves sweeping across the land that are strong enough and that are fluxing at a low enough frequency to excite enough electrical pressure and flow in wires to blow fuses (as is alleged in the Hawaiian incident).

When the nuclear blast ultimately causes these EM waves to arrive in our neighborhood, they arrive from a certain direction. They sweep across long-wire (or other) circuits. All electrical circuits are in loops. That is so electrons can flow in a circle around the loop. The word "circuit" means circle or loop in electrical terminology. Even power lines that appear to be in lines are actually configured in loops with different wires carrying current in opposing directions on the same poles or with one wire carrying current in a loop back to its source.

If EM waves arriving from one side sweep across this loop, they are sweeping down one side of the loop and up the other at the same time because the loop encompasses a total of 360 degrees. Even if the circuit, the loop, is in the shape of a square, a pentagon, a long thin rectangle made largely of parallel wires on power poles, or some random meandering odd shape as in the case of the Hawaiian street lights, it will encompass 360 degrees so that the electrons can get back to their starting point in the wire loop. But, if the EMP directional sweep originates from outside the loop, the EM pushing force would cancel out and there would be no directional effect.

For example: In the case of the strings of Hawaiian street lights that were allegedly impacted by an incoming EM force originating from the direction of the Starfish Prime blast, any pushing force exerted directionally on electrons in one part of the loop would be counteracted and cancelled by an opposing directional sweep pushing the other way on the wire in the loop of street lights. The push on the back side of the loop would cancel any push on the front side. The "push" exerted on electrons on one side of the loop is opposed by a push sweeping down the other side of the loop at the exact same time. There is no circular pressure on electrons in the loop that can cause them to move to fill electron "holes" in other parts of the circuit—resulting in a current. A clockwise EM push would be counteracted by the simultaneous counter-clockwise push on the same circuit.

EM Force Must be Aligned Differently Around Entire Circuit

That is why useful EM fields, that can actually induce a current, are usually oriented circularly with an EM primary coil wound in alignment with another secondary coil or with a rotating electromagnet inside a circular recipient coil. This allows the same pushing angle of the EM field to continually change so that it can present itself to all angles of the secondary recipient coil. It allows electrons to always be pushed by photons through the wire in the same direction. In a nuclear blast, we don't have a circularly aligned or rotating EM field that is producing a consistent push at all angles to the recipient circuit.

Even if you constructed a very long wire, hundreds or thousands of miles long, in the shape of a circle around the entire perimeter of a nuclear blast to create a complete circuit, you wouldn't get an electrical flow. The blast would need to create a massive fluxing low frequency EM force within the loop that sweeps across all parts of the loop in a consistent circular direction. Even if this was postulated from some sort of a supernatural spinning and polarity fluxing nuclear EM field from the blast, it would still be impossible, since this imaginary EM force would be pushing outward towards the distant secondary circuit, the recipient loop, and not sweeping along it in a consistent direction due to the long distance and lack of close alignment. So, it still wouldn't work.

Wire loops, circuits, exist in practically every electrical application (like power lines, street lights, and even in small consumer electronics devices) because the electrical flow needs to loop back to its source either directly by moving from negative to positive or in alternating fashion. Electronic devices cannot function without this circular configuration of the circuit that will allow current flow.

Backwards EM Sweep Cancels Forward EM Sweep

So, any electrical potential induced on one side of the circuit would be cancelled and opposed by a contrary force from the equally oriented EM waves from the direction of the nuclear blast sweeping down the other side at the same time. This would cancel any potential for a directional current flow. This is another serious problem and another strike against the nuclear EMP theory. There would always be a back side to the circuit that is being swept by the same EM waves, but in an opposing direction.

The importance of this issue will become clear when we look at the notorious street light circuits on the island of Oahu at the time of the Starfish Prime nuclear blast, and the tortured logic used to conclude that a nuclear EMP interacted with the street light wiring by sweeping across the lighting loops at a certain angle to the entire circuit.

So, to sum up this issue regarding how EM waves arriving from a specific angle will cancel out when hitting a circuit, we note that the entire affected looping circuit must present itself at an angle suitable to receive a "push" from the waves arriving from the primary EM source if the circuit is to receive an induced current. This consistent push is impossible when considering the action of EM waves sweeping the whole circuit at a specific angle from outside the circuit. For this reason, purpose-built EM transformer circuits have recipient loops of long wire matching up exactly alongside loops of long wire in primary EM coils. Only in this way can the whole loop of wire in the recipient circuit continually present the correct angle to the fluxing EM waves emanating from the primary circuit. This allows electrons to be pushed along a spiraling coil of wire with a correctly aligned push at every changing angle.

10

What Researchers
Want to Find

Electrical Induction in Long Wires

A long metal conductor (e.g. long wire) in the recipient circuit (the circuit developing an induced electrical flow) often assists in effective induction in that circuit from an external EM source. Because of this, circuits (like those in a transformer) that are designed to receive an externally excited flow of electrons in keeping with Faraday's Law of EM Induction involve long wires—often long windings of tens, hundreds, or thousands of feet of wire.

Forecasted Failures in Long Wires

In order to maintain the impression that it is seeking scientifically valid outcomes in keeping with Faraday's Law, the EMP crisis community has historically looked for, and predicted, mass failures in long wire systems from nuclear EMPs. Those wires would allegedly be the perfect lure to draw in electrical forces originating with a nuclear blast. The long wire theory also lends itself to mass hysteria over the long wires in the "power grid."

The theory that wires may be affected electrically by nuclear blasts is not new. Renowned physicist Enrico Fermi considered the possibility of something like a near-field electromagnetic pulse affecting exposed wiring in the very first nuclear detonation in 1945, the

'Trinity" test. He ensured that control wiring was electrically shielded in case such an effect was to occur.

In later years, enhanced theories speculated about a very long distance damaging electrical effect from high altitude nuclear blasts. The blasts had to be high enough to exert this effect upon the Earth's magnetosphere. This elusive creature that can allegedly lash out over a distance of thousands of miles and harm electrical infrastructure is what we currently refer to routinely as an EMP, or an HEMP.

Hence, the two main anecdotal incidents used as examples of alleged EMP damage involve very high altitude blasts and systems connected to long wires. The only one of them that was considered to be documented adequately to be verifiable and studiable, and the only one officially examined in detail by the U.S. Government and the U.S. scientific community, was one in Hawaii involving strings of street-lights connected in series.

There is a reporting of a long-wire underground system having failures in the Soviet Union during one of the many Soviet nuclear bursts over Kazakhstan, but it is almost purely anecdotal and is nebulous and rife with inconsistencies, prompting more questions than answers; like, why didn't all of Kazakhstan have failures in electrical systems since EMP radius calculations said that it should have?

The Hawaiian incident is the focus of this book since it has been officially studied by the U.S. Government and has been branded as authentic nuclear EMP damage. It is the basis for all the EMP nonsense we hear nowadays about the "power grid" in the U.S. The Kazakhstan allegations will also be touched upon.

Random Small Scale Outages Sought on Key Dates

Remember, crisis peddlers want to see EMP effects; so they review occurrences over millions of square miles looking for isolated power outages that occurred during the approximate dates of nuclear blasts, even though no mass nuclear EMP produced failures have ever been observed. So, the EMP pushing community relies almost exclusively on the incident on the island of Oahu for the definitive "proof" of damaging induced voltage and current flow in long-wire systems allegedly stemming from a nuclear EMP.

A thought experiment: If you picked a random date like today's date and spent the next 27 years scouring records and interviewing residents, government employees, and utility company employees over an area of millions of square miles looking for electrical outages so you could validate your pet theory that, let's say, "witchcraft" would cause failures on that date, you would find some sort of minor outages that the various power companies were repairing on that very date. Utility companies are always fixing things—every day. Those minor outages would not confirm your theory that mass electrical infrastructure damage would be forthcoming from "witchcraft." Neither did 27 years of looking for minor outages that occurred on July 8, 1962 allow the U.S. to conclude that a nuclear EMP caused those outages.

11

High-Altitude
Nuclear Explosions

Many High-Altitude Nuclear Tests Were Studied for EMP Effects

The 1962 Starfish Prime nuclear explosion was purposely detonated at high-altitude (250 miles / 400 kilometers / 1,300,000 feet) to observe the various effects, with the primary focus being on the electromagnetic effects. Allegedly, higher altitude detonations produce bigger EMPs that cover more territory. There were other nuclear detonations in outer space, but Starfish Prime was the largest. This is the very type of high-altitude, high-yield explosion that EMP advocates fear the most. It is in a megaton yield category using a sophisticated launch vehicle; neither of which nations with emerging nuclear capability are likely to possess.

Even in its nuclear heyday, awash with funding for bombs, launch vehicles, and staff, the U.S. had a hard-time bringing about this high-altitude megaton burst, aborting similar failed attempts in a run-up to the successful one. That being said, even this device, the scariest of the scariest (for EMP purposes) didn't produce the EMP effect that is claimed. But, Starfish Prime wasn't the only high-altitude nuclear event that would have produced widespread EMP effects, if such things were real.

Starfish Prime was part of a series of high-altitude nuclear bursts over the Pacific Ocean. The U.S. had already tested approximately 278 nuclear devices before Starfish Prime. It has tested a lot more since then. None of them produced mass electrical outage EMP effects. As of this writing, the U.S. has conducted a total of 1,132 nuclear detonations. The total number of worldwide nuclear detonations is approximately 2,476. No mass EMP damage has ever occurred from any of the blasts. Yet, incredibly, we are still required to tremble at the thought of a nuclear EMP.

In the high-altitude nuclear tests conducted by the U.S. and the U.S.S.R., nuclear devices were carried aloft by rockets to various altitudes. Numerous other atmospheric blasts were effected via aircraft delivery vehicles. High-altitude nuclear bursts conducted by the U.S. and the U.S.S.R. that were studied extensively because they were considered to be unquestioningly high enough (above 12 km per their theories) to produce mass-damage EMP effects include:

USA—Yucca; 1.7 kiloton detonation in 1958 at 26 km altitude

USA—Teak; 3.8 megaton detonation in 1958 at 76 km altitude

USA—Orange; 3.8 megaton detonation in 1958 at 43 km altitude

USA—Argus I; 1.7 kiloton detonation in 1958 at 200 km altitude

USA—Argus II; 1.7 kiloton detonation in 1958 at 240 km altitude

USA—Argus III; 1.7 kiloton detonation in 1958 at 540 km altitude (The highest known man-made nuclear explosion)

USSR—Test #88; 10.5 kiloton detonation in 1961 at 22.7 km altitude

USSR—Test #115; 40 kiloton detonation in 1961 at 41.3 km altitude

USSR—Test #127; 1.2 kiloton detonation in 1961 at 150 km altitude

USSR—Test #128; 1.2 kiloton detonation in 1961 at 300 km altitude

USA—Bluegill Triple Prime; 410 kiloton detonation in 1962 at 50 km altitude

USA—Starfish Prime; 1.4 megaton detonation in 1962 at 400 km altitude (The largest man-made nuclear explosion in outer space; the one most likely—per "experts"—to produce widespread EMP damage effects; and the object of the official EMP study discussed in this book)

USA—Checkmate; 7 kiloton detonation in 1962 at 147 km altitude

USA—Kingfish; 410 kiloton detonation in 1962 at 97 km altitude

USSR—Test #184; 300 kiloton detonation in 1962 at 290 km altitude

USSR—Test #187; 300 kiloton detonation in 1962 at 150 km altitude

USSR—Test #195; 300 kiloton detonation in 1962 at 59 km altitude

As you can see, there were quite a few high-altitude nuclear tests that would have produced observable wide-scale EMP damage to electrical systems if such was to be forthcoming. There were also many atmospheric tests at various altitudes that are not listed above. Perpetuating fear by proving widespread damage to civilian systems is key to obtaining funding for EMP research. Civilians pay the taxes, so civilians must be in fear. The above Soviet tests were all over land, over Kazakhstan, and should have produced massive widespread EMP damage, if such was to be forthcoming. More on the Soviet tests later.

12

Starfish Prime

Starfish Prime Nuclear "EMP"

One of the rationales for the Starfish Prime test was that previous high-altitude nuclear tests did not involve a massive enough array of instrumentation over a wide enough area and enough collection of EM and other data that was desired. No significant EM effects had been noted from other blasts, but they should have been there, per their theories. So, this was the granddaddy of them all; lots of monitoring equipment all over the place gathering lots of data. It was years in the making and the U.S. was expecting to learn a lot. The problem is that, from an EMP-production perspective, it was a flop.

The Hawaiian street light incident, allegedly caused by a Starfish Prime EMP, is still dredged up and discussed to this day with a lot of interjected fear as the only definitive example of civilian electrical infrastructure damage from a nuclear EMP that can be turned to with authority. That incident is cited continually in EMP literature as proving the phenomenon. The "proof" of the EMP damage is in the form of a much revered U.S. funded study that I will show to be absolute hogwash.

Brilliant Flash Turned Hawaii's Night into Day for Seven Minutes

What was the Starfish nuclear blast like, as observed from Hawaii? A newspaper account in the July 9, 1962 Honolulu Advertiser described it as a "brilliant flash that turned Hawaii's night into day with a spectacular pyrotechnic aftermath lasting for seven minutes. It was like turning on all the lights all over the Hawaiian Islands for a super-super athletic contest."

The July 20, 1962 edition of Life Magazine showed a late-night crowd on a packed Waikiki Beach gaping at the bright sky. Life Magazine correspondent Thomas Thompson said that the night sky in Hawaii became "brighter than noon."

Well, the blast was definitely visible to the naked eye from Hawaii. More than just visible. It produced a daytime level of brightness per observers. Visible light is part of the electromagnetic spectrum. So, the readily visible display made it clear that "EM" energy reached the Hawaiian Islands from the blast. Let's see what EMP damage was supposedly inflicted on the electrical infrastructure of Hawaii, making it so deserving of scrutiny. If we can't get an EMP effect out of this high-altitude megaton burst, then all hope is lost for the fear mongers that say the whole continental U.S. can be knocked out with an EMP from a single nuclear explosion.

Hawaiian Street Light Incident

The "definitive" study of the Hawaiian streetlight incident was funded by the Department of Energy and was completed by the Electromagnetic Applications Division of the Sandia National Laboratories in Albuquerque, New Mexico in 1989. It is titled, "Did High-Altitude EMP Cause the Hawaiian Streetlight Incident?" This is the study supposedly proving that streetlights went out in Hawaii from a nuclear EMP although it reads like a paper debunking the theory. Maybe I'm one of the only ones that ever read the body of the report instead of just the conclusion. As to bias, remember that this was written by guys funded by the government who wanted a nuclear EMP threat to be real.

U.S. Government Fear: "EMP Would Lose Stature as a Threat"

Why did the feds need to study this incident? Here, in their very words, from their official study, is their reason:

> The streetlight incident is repeatedly quoted in EMP
> reviews and seems to have originated with a report

by S. Glasstone and P.J. Dolan who assert: 'One of the best authenticated cases was the simultaneous failure of 30 strings (series-connected loops) of street lights at various locations on the Hawaiian island of Oahu' ... With the attention surrounding the EMP threat over the last few years, one would expect that detailed analysis of available interaction data would be forthcoming; yet such details have been largely neglected. Even Glasstone and Dolan omit references or a basis for this assertion.

So, Sandia starts out by acknowledging what I have also noted in the first chapter of this book, that the "streetlight incident is repeatedly quoted in EMP reviews" and that there is an ongoing "assertion," with no "basis for this assertion," that the alleged incident, which occurred 27 years previous to the study, was from an EMP, even though there was never a "detailed analysis" of that claim. Sandia continues by describing the small nature of the incident and pointing out that the small percentage failure is not indicative of EMP mass destruction:

> M. Rabinowitz pointed out that even 30 strings of streetlights is a small percentage of the total number in the system and are **not indicative of mass destruction of electrical systems**. There have been suggestions that the **30 strings of streetlights is an overestimate of the EMP damage**. [Bold emphasis added.]

Sandia then appropriately perceives that the small nature of the incident may harm the threat status of EMP:

> If the EMP interaction with the lights was too small to have caused damage, the best authenticated case would be dismissed and the **EMP would lose stature as a threat**. If the EMP interaction was large, then all the lights would have been damaged. **Since that did not occur, the EMP would again lose stature as a threat.** Because of these rather narrow limits and the lack of many data points, because of the wide

> publicity attained, and because of the diversity in views concerning EMP, **the streetlight incident is a crucial verification of the EMP threat to civilian systems.** [Bold emphasis added.]

So, this was their only chance. They had to get an EMP threat out of this incident. It was now or never. There were never any widespread damage incidents from a nuclear EMP to study, so this tiny event was it. EMP could not "lose stature as a threat" and this 27-year-old (at the time of the study) alleged minor incident was the only way to keep the "threat" alive. As of the writing of this book, it has been 55 years since this "best authenticated" EMP damage event. To this date, the 1989 report still stands as the unassailable, although solitary, scientific proof "by the experts" that nuclear EMP damage is a significant threat to civilian electrical infrastructure.

13

Official Study
of the
"Definitive" EMP Event

At the beginning of the official report of the 1989 study, it is made clear that no attempt to gather data for the specific purpose of reviewing the possible EMP causes of the streetlight incident was made until 1985 when John Mattox of Stanford University travelled to Hawaii to try to gather information to prove EMP causes for the incident for a separate presentation he was making regarding the "EMP Threat" at the 1985 American Physical Society's "Forum on Physics and Society." That was 23 years after the incident.

All Records Destroyed Many Years Before Sandia Study

The official Sandia Laboratories report also says that the Honolulu City and County Street Lighting Department destroyed, per their own policy, all records after 10 years, which resulted in a situation where any and all repair records, damage reports, or descriptions of normal day-to-day operations of the utility company were destroyed at least 13 years before any attempt was made to study this presumed nuclear EMP event. Up to that point, it had all been taken at face value as reported by multiple outlets: Yes, of course, a nuclear EMP caused extensive streetlight failures on the island of Oahu. And remember, this is the one and only "well-documented" example of such an EMP event from the U.S. perspective; and the one that the "experts" rely on to prove the existence of the threat.

Sandia Research Method: Phone Two Guys, 27 Years Too Late

So, how did the Sandia National Laboratories use their U.S. Department of Energy funding to conduct the definitive study of this supposed EMP event when there were no records to review? They talked to two guys on the phone who had worked previously in Hawaii in 1962 who gave them some unprovable anecdotes. Only one of them, F. William Souza, had actually worked in the Street Lighting Department. He was the same guy who talked to Mattox in 1985. He was employed by the Honolulu City and County Street Lighting Department on that fateful date in 1962—which was 27 years (!) before the phone interviews for the definitive 1989 EMP study. He repeated and expanded upon anecdotes that he first relayed to Mattox in 1985. There were no records available, so unproven anecdotes became the basis for the report.

Affected Oahu Street Light System Non-Existent at Time of Study in 1980s

They talked to a second man (Alan S. Lloyd) on the phone who used to be a "sales engineer" for the Honolulu Electrical Products Company at the time of the blast in 1962. He *did not* work for the Street Lighting Department in 1962. They talked to Lloyd (who later was a staff engineer at the Hawaiian Electric Company when the interviews were conducted in the 1980s) because Mattox from Stanford had talked to him when doing his on-site visit to gather data in 1985. The Sandia team also reviewed data related to historic fuse and transformer samples obtained in the 1980s that, they admit, may or may not match the types in place in 1962 since the street lighting system and specifications had totally changed over that time period and there were no records from 1962 to review.

Tiny Failure - Much Less Than 1%

So, what is the worst case scenario for damage to the street lights if we accept all the presumptions and all the anecdotes gathered for the Sandia National Laboratory report? The official report refers to newspaper reporting from 1962 that indicates that 30 strings of series-connected street-lights failed on the night of the blast in 1962. How many series-connected strings of lights were there? 3,000.

The report clearly says that the failure was of only 1% of the entire number of series-connected street light strings. And the 1% failure rate doesn't even address the fact that non-series street lights in Honolulu were the predominant type and had no failures. So, the failure, when looking at all street lights, was much less than 1%, at most 0.5 %. This also doesn't include the fact that other systems (power, phone, and communications) didn't fail. So, we are talking about an electrical system failure rate (when looking at all electrical systems in Honolulu) of much, much, smaller than 1%; possibly 1/100 of 1%.

The report also says that, anecdotally, 4 of the series-wired strings would go out on average during storms. If you dig further into the report, you will discover that Sandia says that 23 strings (as opposed to 30) were the actual number of failures on the night of the Starfish Prime blast according to the ascertainable information and the former employee; although he said it could be higher, if the employees didn't bother to report every one. Incidentally, that could be said about any night. The failures on any night could always be higher if they weren't all reported.

So, about 4 strings (anecdotally – no records) go out during storms which amounts to less than 1% of the 3,000 strings on the island, and 23 strings, per the report, went out on the night of the nuclear event (once again, no records to verify) which also amounts to less than 1% of the total strings. And remember, the introduction to the report itself says that the number of affected strings may be an overestimate.

EMP Conclusion Unwarranted from Very Small Failure Rate

We could stop right here and say "case closed."

This tiny failure rate could not possibly give us the basis for a grand theory of mass destruction. But, the statistical insignificance of any failures is just the tip of the iceberg of strong evidence *against* EMP contained in this official report.

Yet, if we jump straight to the official punch-line, this very U.S. report ultimately proclaims that "Evidence indicates that the damage was EMP-generated" and that the official analysis "supports earlier claims of EMP effects on the Hawaiian streetlights." Sandia also parrots and rubber stamps EMP researcher John Mattox and re-states his conclusion that the 1% failure was "definitely associated with the Starfish explosion."

14

Troubling Facts About
Honolulu Street Lights in 1962

This section describes aspects of the Honolulu street lights that made them failure prone from causes other than a nuclear EMP. Normal occurrences would easily explain a small number of outages on any particular day. The below information regarding existing problems within the Honolulu street light system was included in the official Sandia report.

Varying High Voltages Caused Confusion During System Conversion

At the time of the blast in 1962, Honolulu was converting over to a higher voltage (12,000 volts) to supply power to the street lighting systems. However, both types of high voltage (12000 / 7200 volt and the original 4000 / 2400 volt) were present and fed parts of the street lighting system at the time of the Starfish blast. The 3,000 strings of series-connected incandescent lights were present in two-thirds of Honolulu. They were always run as a secondary circuit off the predominant mercury-vapor lighting system which had no failures on the night of the nuclear event. Voltages were typically segregated on lighting poles with the higher voltages appearing higher up on the poles.

The study could not ascertain how the new higher voltage (12 kilovolt) feed lines were separated from the older predominant feeds operating at 4,000 / 2,400 volts for most of the city. It appears that word of mouth or some informal knowledge amongst workers differentiated the two types of high voltage that fed into the transformers feeding the street lights. The report comments on that subject:

> The electric power community refers to this incoming line as a 4-kV line. However, we will call this incoming line a "2400-V" line and reserve "4000 V" to refer to the line so labeled in Figure 3. ... Wouter's notes indicate that **the wiring scheme was unusual**. The electric utility had converted many of their distribution circuits (2400 V phase to neutral in Figure 3) to 12 kV (7200 V phase to neutral) to meet increased power demands. However, the City and County of Honolulu was gradually phasing out its 6.6-A series systems and **did not want to purchase new 30-kW constant-current regulators with 7200-V primaries**. The utility provided 7200-V to 2400-V step-down transformers so that the existing 30-kW regulators could remain in service until the 6.6 A series systems were eliminated. Few had been eliminated by July 1962. [July 1962 was the month of the Starfish prime blast. Bold emphasis added.]

Popsicle Sticks Routinely Used to Hold Contacts Open on High Voltage Transformers

Contributing to the confusion (and likelihood for failure from non-EMP reasons) was the verification by Sandia that odd practices were routinely implemented by the street lighting employees, like the regular use of popsicle sticks to hold the contacts open on high voltage transformers so that lines could be connected to the open high voltage contacts and then dropped lower on the poles to make fuses more accessible for changing. This possibly violated the usual voltage segregation on the poles, but made it easier for employees to fix the fuses that failed often enough in the series-lighting system to require regular repairs. It is likely that the unknown configuration of the popsicle sticks (a non-standard electrical device) and the accepted use of such non-standard practices could produce negative consequences.

Obviously, the confusing mixed high voltages operating in a hybrid system of series and parallel lighting using informal fusing solutions created the potential for excess voltage to accidentally be connected to or shunted to lower voltage systems. This would increase the likelihood for failures from things like blown fuses.

Voltage Varied with Number of Lights on Each String

The Sandia report also indicates that "The voltages varied somewhat, depending on the number of streetlights in the circuit." So, even between each string of series-wired street lights, the voltages and loads were different depending on how many lights had been added to that particular circuit to light each random-sized neighborhood. This practice was destined to cause a lot of failures in the series circuits since specific circuits had different fusing, transformer, and load characteristics. This probably resulted in load and voltage "guesstimations" when employees determined transformer and fusing sizes needed to get the individual lighting circuits to function. This is proven by the Mattox and Sandia observation that normally replaced fuses sometimes failed at the moment of replacement causing employees to insert more plastic washers into the fuse to override the rated failure voltage of the fuse.

Series lighting circuits require different voltage depending on the number of lights—each light adding cumulatively to the required voltage. This is not so with standard parallel lighting or with any parallel wired appliances or devices. This is a problem unique to series circuits; which is why series circuits are no longer used for standard home wiring, street-lighting, etc. For example: 20 lights rated at 25 volts each (like the Hawaiian series lights) require a series circuit voltage of 500 volts; 25 series-wired lights rated at the same 25 volts each require a circuit voltage of 625 volts; 30 lights require 750 volts; etc.

Confusing Mix of Transformers

The report also mentioned that a confusing mix of transformers (4 kVA, 5 kVA, 7.5 kVA, and 10 kVA) was used to deal with various levels of voltage and load since the voltages and line loads were not standard across the lighting systems in Honolulu. Each size transformer had an associated burn-through voltage for lead / plastic cutout type fuses. The Sandia report indicated discrepancies on reported burn-though voltages. It also mentioned that employees would routinely stack on additional plastic washers when lead (Pb) "cutouts" in the fusing system burned out regularly. This informal stacking of additional plastic washers would be an indication that circuits were running at voltages exceeding their design parameters putting them closer to serious failure levels and making them more likely to fail during standard usage than other systems like the mercury vapor lighting system. The official Sandia report also mentions discrepancies in fusing / transformer descriptions from the public utility making it impossible to discern what arrangement was present on any particular circuit.

One section, excerpted below, shows the frustration of the scientists at Sandia Laboratories (in the wake of their discovery of the popsicle-stick system for holding transformer contacts open and the varying levels of high voltage) who were trying to comprehend the hodge-podge of fusing / transformer solutions deployed by the utility company in Honolulu. Neither the public utility nor its employees seemed to care much about proper circuit design or adherence to standard circuit protection protocols since they had no standard voltage cut-off values for fuses. Sandia:

> C.N. Vittitoe obtained samples of this type of disk cutout [lead / plastic fusing system] from Masao Bentosino of the Streetlight Department. ... The average breakdown voltage was listed as 2,700 V. **The breakdown-voltage from Table 2 is not understood.** It likely results from a change in cutout design. The listed values are handbook values. **It is likely that the spread in burn-though voltage is wider after several repair operations.** [Bracketed comment and bold emphasis added.]

Table 2 being referred to in the Sandia quote above shows a 1,500 to 2,100 volt burn-through voltage for the fuses instead of the anticipated 2,700. This put Sandia into a further quandary about the specifications of the system and the voltages needed to cause a fuse failure during day-to-day operation. The "cutouts" being referred to were portions of the fusing system for the strings of street lights consisting of lead disks layered with plastic washers. Were the fuses actually too small for the anticipated loads as seems to be indicated in the referenced Table 2? If so, that would result in fuse failures during normal operation even at acceptable voltages.

Regular Fuse Modifications Masked Inappropriate Line Voltages

The lead cutouts (fuses) were repaired and reused after each arcing or burn-through event as part of normal procedure. Repair involved the use of a file to scrape the lead disks by hand and the addition of more layers of plastic washers as needed to prevent burn-through at the operating voltage for each particular string of lights.

The Sandia report evidences that procedure by saying that the lead based "cutout" fuses would sometimes even fail when replaced normally, at which point a "filing might be repeated and an additional washer added."

Addition of plastic washers was done to increase the gap to slow the melting of lead fuses by increasing the failure voltage by unknown amounts. This increased the fusing voltage of the line which, in those cases of regularly blowing fuses, indicated that those lines were either over-powered, over-burdened, or under-fused. The solution deployed by the utility company employees? Ignore the voltage or load issues and add another plastic washer layer to the fuse to increase the fusing gap and failure voltage.

This continual modification of the lead / plastic cutouts and washers resulted in a continual change and wide variation of burn-through voltages in the fusing system used for the street-light circuits. It consequently raised doubts as to the actual operating voltage of each separate circuit which received more stacked washers in the fusing system until the lead disks were spaced far enough apart to stop the circuit-protecting, lead-melting arcs they were designed to initiate.

Sandia rightly saw this problem when they concluded that, "It is likely that the spread in burn-through voltage is wider after several repair operations." In other words, fuses in the system developed a wider range of failure voltages over time from any specified in the standard charts for those fuse types.

But, Sandia failed to conclude, or even consider, that any of these issues could be the likely culprit for the very small number of purported fuse failures in strings of street lights on July 8, 1962.

Variety of Light and Fuse Sizes

Besides the issue of continual hand modifications to fuses described above, there were a variety of lead cutout sizes that were used and chosen based on circuit voltage, number of lamps in the circuit, and size of lamps. Not only did the fuse cutouts vary in size, but they varied by manufacturer. Sometimes the utility company bought various sizes of General Electric cutouts. Sometimes they bought various sizes of Westinghouse cutouts. The report indicating that, "The choice of brand usually varied with the sale price when the need arose."

The report indicates that various size lamps were also used in the various series circuits with sizes varying between 2500 lumens, 4000 lumens, 6000 lumens, and 10000 lumens; each one imposing a different load necessitating the correct combination of fuses, transformers, voltage, and number of bulbs on that string. The possibilities for confusion and system failure were mind-boggling, especially considering that the whole system had partially switched over to a different voltage and that a mix of on-hand components was used to make the strings function.

Multiple Problems in Hawaiian Lighting Circuits Add Up

The confirmed existence of various levels of high voltage feeding the streetlights, the distribution of those various levels of high voltage to a plethora of random size transformers, the failure of the company to procure appropriate high voltage current regulators on the island, the varying voltage in each series string due to differing numbers of varying-size lights, and the exacerbation of the voltage vagaries by the use of a stackable non-specific fusing system made it highly probable that some strings would perpetually fail by operating at voltages that exceeded their design (or by the deployment of fusing solutions that underestimated the unknown loads).

This is especially true in failure-prone series lighting circuits. It is highly probable that some strings would regularly fail from blown fuses, due to the varying informal structure of the system which presented a variety of voltages and loads that likely exceeded the capabilities of some of the circuits. And from the discussion above on fuses, we can see that some circuits in Honolulu were likely under-fused or over-fused since fusing modifications were done on-site creating a wider unknown range of failure voltages—either too low or too high. These very problems were identified in the Sandia report which is the official definitive U.S. investigation of the "definitive" EMP event.

It is somewhat amazing that more failures didn't happen during the normal operation of this street light system. Maybe they did and they weren't reported since the former employee said that failures might not be documented by the employees making the repairs. Maybe they were reported, but we can't tell, since all the records for that era were destroyed long before the Sandia study. And certainly nobody cared to retain any information "for posterity" on burned-out fuses that pertained to other non-EMP dates.

Why isn't anyone demanding to see or reconstruct the repair information for July 7, 1962, the day before the Starfish Prime blast, so that we can make a comparison and determine "what is normal?" We could then speculate about average number of failures, causes, and under-reporting for that other date. For a study to be real, doesn't the sample data (which in this case is non-existent) need to be compared to a control group (also non-existent) to determine what is normal and what is an aberration?

We certainly don't need EMP to explain a tiny number of alleged failures within such a poorly managed system, about which, there are no retained records.

15

1985 On-Site Study

Attempted Study of One Light Circuit

Mattox concluded at the time of his study (23 years after the Starfish event) that "It was not possible to identify the particular lighting circuits that failed back in 1962. Official records were kept for 10 years or less."

Despite the lack of records, Mattox wanted to physically look at a representative location of a failed string. He set out to try to find the location of a particular string of lights (mentioned in an old newspaper account and in notes from a previous data assessment) that had purportedly gone out in 1962 from a blown fuse. He wanted to make an actual physical assessment of the location to glean information about its configuration. Considering the decades of manufactured hype about EMP damage, this attempt to ascertain facts was commendable, even if it didn't give him the result he sought. Sandia also admitted that its data assessment and 1989 report conclusions were largely repeats of Mattox' work based on his 1985 site visit, so his observations were important to Sandia in 1989 as well.

Mattox Studies Ferdinand Street

Mattox focused in on Ferdinand Street as being the probable location of one of the strings that failed. He couldn't confirm, however, the looping path of the string through the neighborhood or how the connections had been made within the string of lights to the meandering side streets that traversed Ferdinand. Those side streets had been lit by the same string that lit Ferdinand in 1962. It will assist the reader to note that this type of lighting and associated wiring, transformer, and fusing systems were long gone when Mattox visited Honolulu in 1985.

Mattox looked at intersections in that neighborhood and speculated about the possible wiring from decades past. He concluded that the wiring connections and the route of the wires could not be determined on the Ferdinand string of lights.

He had with him a partial historical diagram for lights in that neighborhood, but he couldn't determine the specific route, connections, direction, and orientation of the wires. He gave the intersection of Ferdinand and Puuhonua as an example of his confusion. Per his diagram, Puuhonua (an angular side street) had been lit by the same series string of lights that lit Ferdinand and other neighboring streets. This inability to determine the wiring orientation is important since the azimuth angle of the strings of wires was considered important by Mattox and the Sandia team to determine whether an EMP travelling on a specific heading from the Starfish blast location could induce an electron flow in the street light wires via electromagnetic action. Mattox:

> Wiring details are ambiguous at intersections such as where Puuhonua and Ferdinand come together in Figure 2. The wire with the streetlights on Ferdinand Street may have been connected to the wire with streetlights on Puuhonua Street, or it may have been connected to the wire that continued down Ferdinand Street.

Indeterminate Azimuth Angle for Only String That Was Studied

This no doubt frustrated him because the main goal of his visit was to determine the "azimuth angle" of the wires in relation to the direction of the Starfish nuclear blast to see if they were aligned at an appropriate angle to receive an EM induced current from the blast (which I see as a scientifically impossible proposition as described previously; but nonetheless that was his mission—at which we can see, he failed).

He was unable to make a determination of the azimuth angles for the various portions of the string of wire (theoretical EMP recipient) on the streets in the neighborhood since the older series lighting system no longer existed. It was gone, as were the records. He was only left with vague speculation as his guide.

This is important. This solitary on-site physical study, despite its lateness, should be given more weight than the many pseudo-scientific attempts to place imaginary data into computational models (seen later in the Sandia report) to infer the truth or fallacy of a nuclear EMP in this situation. It was the only actual attempt in both the Mattox and Sandia study to get an empirical grasp on the structure of the actual wiring, the circuit specifications, and the azimuth angles of the wires for a failed string of lights. But, it was a total failure since the string of lights no longer existed and its former path was indeterminate. He could not therefore extrapolate any estimates about the wire orientations for the other failed light strings in other areas, either.

Everything else in the study, which is built up from a presumption of angles, is just a whiteboard think-tank exercise devoid of observations and facts.

This would be another opportune moment to say, "Case closed" on the nuclear EMP theory, but we continue.

16

Sandia Ignores Facts

Sandia Concludes Correct Azimuth Must Have Existed

The orientation of the wiring in the strings of lights became important in the Sandia study, largely because of the speculation that the presentation angle of the wires in relation to the supposed incoming nuclear EMP was crucial for the prospect of proving an inductive coupling in the failed strings. This was made problematic when it was discovered from discussions with a former employee that strings of lights wandered around blocks and back-and-forth across streets. This, in conjunction with Mattox' indeterminate assessment from his on-site inspection and the lack of records made it impossible for the Sandia team to determine the former paths and angles of the now non-existent strings of series lights that allegedly failed.

Yet, ultimately, the desired conclusion was decreed despite the inability to verify the various angles of the wires and the phase cancelling that would occur when wires present angles harmful to incoming waves or when ground reflections or other reflections impact the same circuit.

Ancient Anecdotes Substitute for Facts

Mattox' lack of records to study or systems to analyze led to the introduction of anecdotal evidence that was re-used by the Sandia Team.

Anecdotal evidence (whether it pertains to Bigfoot, space aliens, criminal conspiracies, or EMPs) has a tendency to expand over time to fit the narrative desired by interviewers. Reliance on very late anecdotes, when faced with a lack of real evidence, can be especially risky when there is only one source, a solitary "witness;" in this case, a former employee. This solitary witness became very handy when the Sandia team needed random facts to nail down their theories.

Multiple Problems

Clearly, the hodge-podge status of the street lighting circuits was problematic and introduced a high chance for regular circuit failures from voltage / load / fusing discrepancies and from individual bulb failures in series strings which would cause the affected string/s to go dark. Budget concerns apparently precluded the procurement of the correct array of current regulators and transformers to deal with the multiple levels of high voltage criss-crossing the city. You don't have to be an electrical engineer to see the problems and potential for the tiny number of fuse failures that occurred on July 8, 1962 (or on any other day).

Lighting System Had No Failures on All Primary Circuits

Here is another serious problem with the EMP failure theory: Every one of the thousands of series lighting strings was fed off of a primary string of lights that had no circuit failures.

The series lighting circuits for incandescent street lights were fed as a secondary circuit off of a high voltage lighting circuit (2400V / 4000V) powering a parallel-wired string of mercury vapor lights. Various sizes of step-down transformers on the mercury-vapor lighting circuits provided varying voltages (500-750 volts, dependent on the number of lights) for the series circuits of incandescent lights; each bulb on those strings operating at 25 volts.

Why didn't the long wires for those mercury vapor circuits (the predominant lighting system for the city) receive damaging EMP induced voltages? If some of the angles were appropriate to induce extreme EMP voltages on the thousands of secondary lighting circuits, why weren't some of the angles appropriate on the thousands of primary lighting circuits that were always present in the same areas as the series strings?

Confusing Patchwork of Systems

These systems existed as a confusing patchwork in the same neighborhoods: Two different lighting systems side-by-side in the city of Honolulu with the primary circuits powering the secondary circuits; one using high voltage with lamps wired in parallel; the other using low voltage lamps wired in series. The power was introduced via an array of varying size transformers and varying size fuses put in place to accommodate a varying number of bulbs and variety of voltages.

Third-World Methodology

It is very reminiscent of third-world power and lighting systems. Having lived most of my life in and near third-world countries, I have often seen those very types of voltage and circuit conflicts in those places. It amounts to a strategy of doing whatever it takes to make the lights come on most of the time despite the lack of appropriate parts, infrastructure, circuit design, internal policies, or training.

Based on my review of the subject and the observations of the Sandia team, I can only conclude that the island of Oahu had, in 1962, what amounted to a third-world strategy for circuit functionality in the street lighting systems. Changes in the system, when they occurred, were sometimes done without the procurement and installation of proper transformers, fuses, and current-limiting devices. Series circuits were made longer and shorter with more or less bulbs (of various lumen ratings) to fit the shape of each neighborhood without full consideration as to the ramifications to the overall load (voltage and current) on that circuit.

Remember, the data and corresponding conclusions on hodge-podge circuits I am presenting here are *from the officially commissioned study*. I am not pulling these things out of the air. Even the persons making the official study found the circuit protocols to be mind-boggling.

Don't try too hard to understand the electrical aspects of the following quoted description, but read this excerpt from the official study just to see a broader glimpse of the mess the lighting system was in and the hopeless task of determining what made a small number of circuits (1% of a secondary system) fail on a date 27 years prior to the study. This description from the study tries to decipher the four voltage categories present on a typical pole and the differing voltage characteristics of differing adjacent lighting systems. This is only a discussion of lighting wires and doesn't even include home / business power lines, police call box lines, fire alarm lines, cable TV lines, and telephone lines on the same poles, which, despite being "EMP-prone" long-wire systems themselves, did not fail. Sandia:

> Figure 4 shows a typical pole arrangement. The 4000-V and 500-V lines have been owned and operated by the City and County of Honolulu. One string of lights is described as perhaps covering a linear distance of approximately 5 city blocks, but at times circling blocks or meandering back and forth.
>
> The 4000-V lines typically had approximately 28 mercury vapor lamps plus perhaps two isolation transformers driving the incandescent lamp secondary circuits of the type that failed. Each 400-W mercury vapor lamp required approximately 1 kW driving capacity to provide the needed starting surge. The secondary isolated circuits were used at side streets characteristic of residential neighborhoods where short poles left no room for the 4000-V lighting systems. The only place where the 4000-V and 500-V lines had the arrangement in Figure 4 is where the isolation transformer made the transition from the primary to the secondary lamp circuit. The 500-V line proceeded from this origin to a series of shorter poles. Alan Lloyd indicated that the uppermost lines were reserved by Hawaiian Electric [at higher voltages]. The City-County lines at 4000 V and 500 V were at smaller heights, as indicated in Figure 4. [Bracketed comment added.]

That excerpt is from a portion of the Sandia study that tries to get a grasp on the configuration of all the wires present on the poles in preparation for the "azimuth" discussion that followed. It was impossible, however, in the late 1980s, to know the operating voltages and fuse set-up of the lighting circuits in each neighborhood and to get a clear picture of wiring layout since the lighting circuits in each neighborhood had been configured differently from each other in 1962, all records had been destroyed, and the now non-existent wires were known to be "at times circling blocks or meandering back and forth."

17

Azimuth Endures as Silver Bullet

Why No "EMP Failures" in Other Long Wire Systems?

As mentioned, Sandia doesn't discuss other non-failing long-wire systems for things like police call boxes, fire alarms, and power lines that are also included in the pole diagram referenced as Figure 4 in the Sandia study. Why not? If they didn't fail, their wire orientation, etc. should be studied to determine whether an EMP effect is believable or nonsensical.

They weren't studied for precisely that reason. They didn't fail and the lack of failure would discredit the theory. It is clear they should have asked the following question: Why would a tiny number of circuits fail in only one long-wire system and not many circuits in all long-wire systems if long wires would, by their very nature, receive damaging induced EMP voltages and currents? That question is never posed and is never answered even though the eventual official conclusion is that EMP damage must relate to the unknown azimuth angles of certain wires in relation to the direction of the blast. That was the theory. Why it would only pertain to one subordinate part of one system is never discussed.

Wiring Angles Could Not be Determined

What is clear from the study is that Mattox and Sandia never got a clear picture of the wiring because of the lack of records and later revamped system; but they tried to. Each neighborhood had a different configuration. The lighting wires weren't configured in a cookie-cutter duplicate design in each neighborhood. Therefore, they could not be studied as a group for azimuth or failure potential from other causes. It became clear to Sandia, and they admitted it, that the former wiring configurations were unknowable in the 1980s.

No Correlation with Azimuth Angle for Nuclear Induced EMP

In order to try to prove theoretically an induced voltage and current, Sandia made up imaginary azimuth angles for the affected wires and placed them in mathematical models to see if they could conjure up wire orientations that would allegedly be more likely to receive an induced electrical force from an EMP coming from the direction of the Starfish Prime blast. This was done since azimuth angle was the silver bullet to "prove" the EMP theory.

Sandia Attempts to Guess Azimuth Angles for Wires

The possible locations of 13 of the failed strings were listed in the Sandia study for analysis because the researchers hoped to determine the wire orientations. What they really wanted was for all the wires to be lined up appropriately to receive an EMP-induced current.

It was documented that the former wires for strings of series lights meandered around entire blocks and criss-crossed streets and went down multiple side streets. This made the azimuth even beyond guessing without records to fall back on. Although Sandia desired an azimuth correlation, the section quoted below regarding orientation of the affected light strings concludes that "such a correlation is not apparent":

> In Table 3 the column for orientation angle has been added to the LLNL data to test for a possible strong correlation with the damage. **Such a correlation is not apparent.** Perhaps knowledge of wiring orientations would show better correlation. Angles at approximately 66 degrees correspond to alignment with a horizontal projection of the line of sight to the burst. Figure 3 suggests that a cross street might orient some wiring at 90 degrees to the listed street,

giving an equivalent avenue of excitation near 24 degrees from north. Seven of the 11 angles in Table 3 are within 10 degrees of these two angles, 66 degrees and 24 degrees. With a random distribution, we would expect about five. [Bold emphasis added.]

So, randomly-chosen wire directions would give five good angles out of eleven. "Good" from their perspective of proving a nuclear EMP. Their guess on wire angles gave seven out of eleven supposedly helpful angles.

Whoop de do. Seven out of eleven based on guesses of the angles compared to five out of eleven from randomly chosen angles. That is not indicative of anything. Remember, there are no good records or existing wires to examine that could reveal the angles for the failed strings. These are only guesses. And even the favorable guesses as to wiring orientation made by Sandia researchers, who want to find an EMP outcome, don't show a correlation. Their official conclusion regarding whether the angles correlate to an azimuth that would induce an EMP voltage spike is, once again, that "such a correlation is not apparent."

18

Sandia Director of Testing Rejects EMP Theory

Dissenter on Sandia Team Denies EMP as the Cause

Another hypothesis made by a Sandia team member (retired Navy Commander Don Shuster, who was the Director of Field Testing at Sandia National Laboratories) rejected the idea of a nuclear EMP as the cause for the small number of failed street lights on the island of Oahu. His hypothesis, although probably more scientifically sound and more plausible than the nuclear EMP theory, was ignored when the team presented its conclusions.

Shuster described a common phenomenon that causes failures in electrical systems. He explained how failures often occur when there is a near simultaneous switching of a large system-wide load from an "off" state to an "on" state. The near simultaneous switching would cause a heavier than normal burden on the system due to the additional start-up power required by all the connected electrical devices. This spike in demand could result in system failures from such things as blown fuses (as was seen in the small number of failed strings of street lights).

Sandia Director Shuster: Photocell Hypothesis

Shuster explained that the streetlights in Honolulu were all activated by photocells. His theory noted that, typically, as night approaches, different areas become sufficiently dark incrementally causing the photocells to activate the street lighting in stages. The street lighting comes on in a naturally staggered fashion since hills, buildings, land contours, and photocell orientation affect the time the circuit comes on. A building or a hillside that shades some photocells more than others would cause different photocells to come on at slightly different times. This is, of course, true.

During a normal evening sunset period, the staggered effect causes street lights to come on over a period of time that could be half-an-hour or more. The activations are not close to simultaneous; not all off and then all on. Therefore, the start-up power demand at sunset is incremental and spread over time. Shuster concluded that the unnatural 11 PM daylight effect produced by the Starfish Prime burst may have caused the circuits to switch off and then, several minutes later, to switch on more in unison than a normal slowly dimming sunset situation.

Recall the newspaper account from the Honolulu Advertiser saying that the entire chain of Hawaiian Islands was lit up like a large athletic event for a seven minute period late at night. Also recall that Life magazine correspondent Thomas Thompson said that the 11 PM night environment in Hawaii became "brighter than noon" from the huge H-Bomb explosion in the sky that dwarfed the Hiroshima blast. Photographer Terry Luke of the Honolulu Star-Bulletin took a beautiful picture shortly after 11 PM that night that is indistinguishable from a daytime picture. It shows the bright blue sky with wispy white clouds, the buildings on the distant skyline, the light blue ocean off in the distance behind the buildings, and the green grass and bushes beside the black asphalt in the foreground.

Massive Start-up Surge at Photocell Activation

A sudden simultaneous activation of many circuits would cause a surge in demand and a large in-rush current that would impact the whole system and easily overload parts of the system causing failures. Remember that each of the mercury vapor lamps (approximately 28 lamps for each primary string that also powered the subordinate "secondary" incandescent series circuit) required a significant surge of energy at start-up.

As quoted previously, "Each 400-W mercury vapor lamp required approximately 1 kW driving capacity to provide the needed starting surge." So, each primary string that also provided the power to the secondary series string via a transformer needed an approximate 28 kW starting surge to light all 28 mercury vapor bulbs when the photocell turned the circuit on. This, per Shuster's theory, was likely happening all over the city and county of Honolulu in a more simultaneous fashion than a typical gradual sunset activation of photocells across the city. A 28 kW starting surge on each primary lighting circuit multiplied times 3,000 for the number of lighting circuits gives us a total surge of 84 megawatts if the photocell activation was simultaneous.

This doesn't include a calculation for the additional power burden put on the entire lighting system from the series strings activating at the same time. Although the power usage varied on the 3,000 series strings due to bulb count, etc., they may have averaged about 5 kW each. That would be another 15 megawatts added to the starting surge for the mercury vapor lamps equating to a total of 99 megawatts. That would be the approximate amount needed to turn the street lighting circuits on across Honolulu in a simultaneous fashion. It is likely that the lighting system would be stressed by such an instantaneous surge in demand. A "near simultaneous" activation would require less power, but still a significant amount. This brief heavy demand could have definitely overburdened the system and caused failures.

Although not causing massive failures, Shuster felt that this surge, to the extent it occurred, could have caused the small number of failures (less than 1%) that allegedly occurred. This theory, presented in the official Sandia report, is very logical and seems to deserve more than an inglorious passing mention followed by an outright rejection. It is no less feasible than the nuclear EMP theory. As explained by Shuster in the official Sandia report:

> The light would automatically turn the streetlights off, then on. At a typical sunset (sunrise) the streetlights come on (off) in a staggered manner dependent on the slow setting (rising) of the sun, on the sensitivity of the photodiode, and on the presence

of shadows. The light on the night of July 8, 1962, was turned off and on much faster than at a typical sunrise and sunset; the synchronized demand to alter many lights at once might have stressed the system and created the failures. A rapid rate of change of current in the 4000-V line, for example, could overstress the 500-V line and allow either onset or extinction of the light to cause the experienced damage.

Sandia Teams Discards Director Shuster's Non-EMP Theory

The Sandia research team (consisting of industry contractors and government personnel) rejected Shuster's hypothesis (or should I say, rejected his attempt to turn off the funding spigot) by merely concluding that an electrical spike, a "transient," caused by simultaneous photocell activation "should be damped by protective equipment."

Of course that "protective equipment" includes fuses and some fuses did blow; those blown fuses being the very issue at hand and the whole purpose of the Sandia study. It seems ludicrous to say that an electrical spike / transient caused by excess simultaneous demand could not cause failures in "protective equipment," but a spike / transient caused by an EMP could. Here are some excerpts from the study, in response to Shuster:

> The peak dI/dt is estimated as the voltage applied by the current regulator divided by the inductance. Any increase in dI/dt because of synchronization of many photocells might increase this voltage. However, it appears that this could happen only by sending the power generator into a transient that should be damped by protective equipment. ... Thus the photocell hypothesis is discarded.

It is ludicrous how such a theory can be "discarded" out of hand by the Sandia team while they, at the same time, uphold a very questionable nuclear EMP theory for a failure that, from the outset of the same report, is viewed as a "small" failure that is "not indicative of mass destruction." They do this while revealing their bias by bemoaning the fact that EMP may "lose stature as a threat" if the right conclusions are not reached. The report clearly articulates a bias against alternate theories when it says early on that "the streetlight incident is a crucial verification of the EMP threat to civilian systems."

Employee Anecdote Bolsters Sandia Against Director Shuster

In order to nail down their rejection of Shuster's photocell theory, the Sandia team asked the one former employee of the Honolulu City and County Lighting Department that they were speaking to about his memories from 27 years ago, if he could recall a close-in-time activation of photocells across the City and County of Honolulu after the blast. He recalled that, at the time, he was watching the nuclear blast from a hillside (as were many others in Hawaii). He then concluded that, from his memory, there was not a near-simultaneous photocell activation. So, that's that.

They now had anecdotal verification that there was no close-in-time photocell activation for 3,000 strings of mercury vapor lights and an equivalent number of series strings on short poles covering many neighborhoods over the entire city and county of Honolulu on the island of Oahu. By the way, the city of Honolulu covers approximately 43,000 acres. Honolulu County (the portions on the island of Oahu) covers approximately 382,000 acres. I'm sure there were lots of palm trees and hills and valleys obstructing the views between neighborhoods too.

The verification was from a guy watching the sky go up in flames from a hillside. He said he didn't notice simultaneous photocell activation happening across the city and county of Honolulu (which he could not possibly have viewed in its entirety, or at all), so the theory was rejected. That's the answer the Sandia team needed to shut down Shuster, so they got it.

"Everyone Awake" Theory

I want to make it clear that I think that the shoddy design and operation of the street light system was the source of the very small number of failures that allegedly occurred on the night when everyone was looking for failures, but while we are discussing alternate theories, I would like to mention another personal theory that is similar to Shuster's photocell theory regarding high electrical demand.

Most people on the entire island of Oahu stayed awake on July 8, 1962 until the 11 PM blast time and remained awake for a while afterwards to see the affects. After all, people wanted to see if their world would come to an end or if their bones would become visible through their hands when a nuclear device 93 times more powerful than the one that destroyed Hiroshima was detonated in the sky. This mass shifting of normal sleep patterns no doubt caused Hawaiians to use more electricity that night. In addition to the street lights, the power system on the island had to supply power for the waking activities of most of the island's population, many of whom were usually asleep at that time of night.

This extra electrical demand on the island's electrical systems was no doubt significant. It was a late-night event that most everyone on the island stayed up for. Increased electrical demands cause increased failures, so a minor (less than 1%) failure of one of the electrical systems connected to the power grid on the island would be in keeping with a theory involving increased electrical demand that night. Widespread workaday waking power usages do not typically occur at the same time as night-time power demands from things like street lights that are on when many people are asleep or involved in more passive low-energy activities.

On this occasion, typical daytime power usages from most people on the island being awake watching TV, cooking food, and using appliances coincided with night time usages (street lights) creating a larger than normal peak electrical demand. Certainly such an increase in demand could explain the very small number of failures (1/100 of 1%?) that occurred in all grid-connected electrical systems when viewed statistically as a whole. These types of high-usage failures happen routinely on hotter days across the U.S. when more people run their air conditioners at the same time increasing the total electrical demand on the system. No one attributes those common minor failures (or even larger demand-driven brownouts and blackouts) to EMPs.

19

Failure Prone Series Circuits

Problems with Series-Wired Lighting Circuits

Series connected lights are notoriously problematic and failure prone. That's why home Christmas lights now primarily use parallel rather than series connections for each bulb. In years past, series-connected strings of Christmas lights failed often because an individual failing bulb would disrupt the whole circuit and cause the whole series string of lights to fail. These same areas in Oahu no longer use failure-prone series lighting. They now use only the common parallel lighting circuits that continue to function despite the failure of individual lights.

Sandia Conveniently Concludes That No Bulbs Were Burned Out on Oahu

To avoid this series-wiring argument stemming from individual failed bulbs, the anecdotal interviews conveniently drew out the "fact" that no bulbs were burned out in the series lighting circuits, only fuses. The 23-year-old anecdotes extracted from the former employee by Mattox conveniently included memories that there were no burned-out bulbs that would have caused entire affected strings to fail. This type of failure is a normal non-EMP occurrence for series-wired strings of lights. The very likely "failed bulb" scenario had to be dealt with head-on, in no uncertain terms, and obliterated to keep EMP on the table.

It is ludicrous to state that 3,000 strings of lights on short poles, each containing a varying number of lights from 20 to 30 (the numbers mentioned in the report) had no burned out light bulbs on any particular night; or that a single employee could know that there were no burned out street lights in the lighting strings across the entire city and county of Honolulu on the night of the blast (which would have resulted in whole series strings going dark if they contained a single burned-out bulb). 3,000 strings of lights multiplied by 20 to 30 lights per string equals 60,000 to 90,000 individual series-connected light bulbs.

The approximate number of bulbs would be 75,000 if we use 25 bulbs per string as an average. This doesn't even include the tens of thousands of mercury vapor light bulbs. How could a former employee possibly know—or dredge up a memory from decades past with no records to fall back on—that none of the 75,000 series-wired bulbs across the city and county of Honolulu were burned out that particular night? It is way more likely that there are *always* some burned-out bulbs when you are talking about a system that contains tens of thousands of bulbs.

A Burned-Out Series Bulb Would Cause Whole String to Fail

By the way, the only actual reporting in existence of the number of lights failing on that night is from anecdotal newspaper accounts like that in the April 8, 1967 (4 years after the event) *Star-Bulletin* which says that "300 city streetlights" went out. Sandia needed failures in each of the small number of failing strings to be from fuse failures, not from an occasional burned-out bulb that would have caused the whole string of lights to fail. So, Sandia and Mattox conveniently got that result by querying the former employee who "remembered" that there were no burned out street light bulbs in Honolulu that night.

How nice. None of the tens of thousands of light bulbs were burned out that night, which, if they had been, would contribute to a non-EMP argument. Failed bulbs would have broken the circuit on affected series-wired strings, causing the failure of the entire string.

Sandia: Voltage Varied on Different Series Strings of Lights

Incandescent bulbs (the type on the strings that failed) are noteworthy for their tolerance of voltage variations burning brighter and dimmer on different voltages, and it seems that the standard was, if the circuit lit up, that was good enough. Voltage consistency and bulb count didn't seem to matter. The flexible hand-modifiable fusing system masked the voltage and impedance inconsistencies in the circuits.

Strings of varying bulb counts, from 20 to 30, are mentioned in the report, the quantities having been selected in each string to fit the meandering shapes of each neighborhood. There seemed to be no standard bulb count, which should have been correctly dictated by voltage. As indicated in the report:

> In some areas the available wire locations on the lighting poles required a reduced voltage from the high values needed for the long series circuits. An isolation transformer at a lower voltage then supplied current for these small loops of about 20 streetlights.

In other words, the series circuit design was flexible and determined by the utility company employee on site. It was flexible as to the number of bulbs, the estimated voltage needed to light those bulbs, and the amount of stackable fusing needed to keep the circuit from blowing. There was definitely no consistency in the street lighting on Oahu in 1962. If it failed, they stacked on more plastic washers to increase the fusing gap between the lead disks. If the lights were too dim, they grabbed a different wire off the pole or put some more voltage into the circuit through a variety of pole-mounted transformers, which differed all over the city.

119

20

"Coupling Analysis" Section of Sandia Report

Garbage In, Garbage Out

After the report fully and admirably states the uncertainties coming out of a lack of records and an inability to observe or ascertain line orientation, lamp count per line, voltage, current, transformer type, and fusing sizes or protocols for each string, it delves into a set of complex calculations reminiscent of what Dr. Sheldon Cooper would put up on a whiteboard on the show, "The Big Bang Theory." But, as the saying goes, "Garbage in, garbage out." The calculations are pointless because they are based on invented data. The study admits that it is impossible to know what the supposedly affected circuits were like or what quantities should be inserted into the Sandia calculations.

"Detailed Calculations Are Not Justified Because of Such Uncertainties"

The first part of the report correctly outlines the problems that prevented the gathering of reliable data. After the data is shown to be totally unavailable, the report shifts to analysis—but, analysis of what? In the section titled "Coupling Analysis," the report tries to explain how an EMP could have acted upon (coupled with) street light wiring to produce a damaging voltage and corresponding current flow in the very small number of lines that failed. Even while playing with and displaying a vast number of calculations to give the appearance of a scientific method, the report says in that same section:

> **Many problem parameters remain a mystery**: surge impedance of the actual isolation transformer, reflection coefficients at each line discontinuity and at the ground, efficiencies for converting common-mode current on the 4000-V line into differential excitation of the 500-V line, geometry for the 4000-V line, geometry for nearby structures and their scattering effects on the incoming wave. **Detailed calculations are not justified because of such uncertainties**. [Bold emphasis added.]

Did you get that? In the very section that is making detailed calculations to prove that an EMP caused the line failures, Sandia says that *"Detailed calculations are not justified because of such uncertainties."* But, of course, the Sandia group couldn't expend their government funds and then conclude that nuclear EMP is not real or not provable from the supposed Oahu Island streetlight event. They do, in essence, say throughout the report that it is unknowable and unprovable, but their bottom line must be something else that is dictated by their paymasters.

Readers are probably getting tired of me saying this, but we could again say "Case closed" based exclusively on this single point. Sandia said that detailed calculations are not justified. They should have ended their study at that point and said, "The cause of the street light incident is indeterminate." The end. Right? But no, we continue into la-la land.

"Imaginary" Data

The analysis section is peppered with admittedly "imaginary" complex calculations based on invented data and tables that are even labelled with the word "imaginary." Language like the following permeates the analysis section, "Consistent with the uncertainties involved, there are several variations that could be examined to alter the open-circuit voltage." In other words, endless variables are plugged into pointless equations to give the appearance of thorough scientific analysis. The deferential language admitting ignorance of the facts apparently forgives the false appearance of reliability given by the hyper-specific calculations. [Note: See "Appendix" at the end of this book for a sample of the ridiculous calculations.]

Regarding the diagrams that are actually labelled as "imaginary" owing to the fact that no data is available for the study: Wouldn't it be convenient if any researcher on any topic could just make up data with corresponding diagrams, tables, and calculations to fit whatever conclusions he desired? It might be an interesting, albeit pointless, exercise to have an engineer do mathematical calculations on the loads and stresses for a high-rise building that he has never seen and has no data for—and then have him present his conclusions as definitive.

Sandia: Damage Should Have Been More Extensive

After the Sandia team plays with enough variables (since real data is not available), they produce a configuration that gives them enough theoretical voltage to clearly blow circuits. Then, the study team correctly asks (after making fictional calculations):

> These results leave one **major contradiction to a convincing argument**. If the induced voltage is so far above the value needed to produce the observed fuse closure, **why wasn't the damage more extensive? Why weren't many more circuits upset?** Rough estimates of RC time constants for the disk cutout (using transmission-line characteristic impedance for R) indicate response times of a few nanoseconds. So the fuses should respond quickly to the propagating signal; response time should not have limited the damage. [Bold emphasis added.]

Very true. They manufactured a workable scenario (from the perspective of validating EMP effects) with invented data and mushy science, but then concluded it should have caused widespread circuit failures. Good point. There were no widespread circuit failures. At least they buried the real truth in the body of the report to be found by those with a slight understanding of physics or electronics who are willing to read it.

Sandia Conclusion: Invented Data Proves Nuclear EMP

They then, conveniently, scale back the invented data in their variables making the values be at the threshold for some damage. Liking these invented numbers better—that don't fit the facts or the science—they conclude that EMP effects are real and that a nuclear EMP caused the Hawaiian street light failures.

It is an odd conclusion. They admit that wire orientation cannot be known. So, capture angles cannot be shown to be adequate (per their flawed theory) to receive an induced voltage due to an EMP originating from the direction of the blast. Other inconvenient factors like phase cancelling from bad angles in the same circuit or reflections cannot be known either; but, if they continually adjust the variables and then plug the right mix into their formulas, they can manufacture the possibility for the very beginnings of a very small number of failures from a small voltage or current variation. Therefore, the nuclear EMP threat is real.

Why does this make it real? All that they have concluded from their armchair view of the event is that a very small number of circuits failed from voltage or current issues. This has always happened in all electrical systems throughout history; especially in hodge-podge systems like the lighting circuits of Honolulu in 1962. Why does this tiny number of failures with no data to support any analysis have to be from a nuclear EMP? They needed an affirmative answer and this avenue riddled with falsehoods gave it to them.

McDonald's Analogy

It's like concluding with no evidence that my average 100 pound son has been obviously overeating at McDonald's because he weighs 101 pounds today. I don't have any data to support this and the one pound (1%) difference is insignificant when considering that it is normal for body weight to fluctuate slightly day-to-day. But, I have some formulas that I will plug numbers into regarding the items on the McDonald's menu. After extensively studying carbohydrate, fat, and protein counts for each menu item, I make complex algebraic calculations and conclude that secret unobserved visits to McDonald's must have occurred that made him one pound heavier.

But a problem arises. After estimating some "good" numbers to support my theory, I conclude that he should be much heavier and not just one pound heavier. So, I crunch the numbers again in my fanciful calculations, this time changing the total grams of fat, protein, and carbohydrates that were ingested and the frequency of the unverified secret visits. After doing so, I still conclude that he must have overeaten at McDonald's, but that he must have just barely eaten enough to show the beginnings of a statistically insignificant trend towards obesity. So, my definitive conclusion is that yes, he absolutely has been overeating at McDonald's and is getting fat, even though I have no proof that he even went to McDonald's. My made-up numbers say it is so. Forget the facts, or lack thereof.

21

Conclusions Section
of Sandia Report

Disclaimers

In the very short half-page "Conclusions" section of the study, disclaimers are made by the apparently guilt-ridden Sandia team:

> Because of the hardware considerations that have been noted, **we cannot claim to have a rigorous analysis of the 1962 Hawaiian streetlight incident**. Unknowns in the complications caused by neighboring wires and lack of circuit details for 1962 prevent a rigorous analysis today. Such data as "How many clear-plastic washers were in the transformer cutouts that failed?" are not now available. [Bold emphasis added.]

It then laughingly continues, "Nevertheless, a consistent view has emerged. That view is supported by the limited analysis."

Really? How can you say both of those contradictory things in a half-page wrap-up? At least they didn't just lie and say that their fabricated data and facts were actually truthful.

They then commit the atrocity of attributing the small (1%) failure within a certain type of street light circuit to a nuclear EMP even though none of the predominant mercury vapor lamp circuits or any

127

of the other long-wire systems in Honolulu failed even to a small extent.

Sandia: EMP is Real

Sandia vaguely concludes in the final section:

> The estimated differential-mode voltages are enough to have caused the observed damage. Orientation effects, expected discontinuities in the transmission lines, and the interaction between differential-mode and common-mode excitations make the present estimates consistent with damage occurring in only a portion of the strings in the system and consistent with none of the individual streetlights being damaged. ... This analysis gives a basis for attributing the cutout damage to an EMP effect and supports earlier claims of EMP effects on the Hawaiian streetlights.

Wow, "estimated" voltages "are enough to have caused the observed damage."

And then with no shown connection:

"This analysis gives a basis for attributing the cutout [fuse] damage to an EMP effect..."

So, they admit to making up voltages that they think would blow a tiny number of fuses. They then use the made-up data to legitimize the Hawaiian streetlight incident as an EMP effect.

Despite this team's laudable moral constraints and efforts to tell the truth about lack of data and shoddy system design in the body of the report, it is obvious that the EMP damage conclusion was a fait accompli. It was decreed and so it must be. The conclusion seems oddly out of place though. Like a movie that had an alternate ending spliced on. I would have preferred the truthful ending; not the one tacked on by the censors.

22

Starfish Prime
Was Thoroughly Studied
For EM Effects

Starfish Was the Most Thoroughly Instrumented Nuclear Blast

A declassified film from the Defense Atomic Support Agency produced in conjunction with the U.S. Department of Energy, Albuquerque Operations Office, describes some aspects of the Starfish Prime test and other high-altitude blasts. The film, known as U.S. Nuclear Test Film #63 and titled "Operation Fishbowl: High Altitude Weapons Effects," describes the wide array of instrumentation platforms and the large geographic area studied for each high-altitude nuclear blast. It also described the various types of electromagnetic effects that were researched.

For Starfish Prime, there were 266 official sea, air, and land instrumentation stations gathering data. This does not include the three instrumentation pods (recovered later for data analysis) attached to the nuclear-tipped launch vehicles in Operation Fishbowl that were deployed directly, at various altitudes. Nor does it include the many instrumented missiles fired into the nuclear fireball during the nuclear detonations. It also does not include the thousands of military communications devices in the Pacific area that were redirected to various EM communications frequencies to participate in the test at the time of the blast to test the effects on different portions of the electromagnetic spectrum. It also does not include satellites like U.S. Explorer XV that were used to study the EM effects of the blasts upon the ionosphere and the artificial radiation belt created by the blasts.

266 Instrumentation Stations

All 266 instrumentation stations were geared to collect data to ascertain the effects of nuclear blasts at high-altitude.

27 instrumented missiles were fired aloft from Johnston Island. There were ships dedicated to the Starfish Prime test that were outfitted with instruments throughout the Pacific. There were test beacons, tracking systems, and radio relay stations assessing EM effects at different angles through and around the blast area. There were dedicated instrumented test ships offshore of Johnston Island and hundreds of ships deployed from many Pacific islands and the mainland U.S. whose missions were temporarily redirected during the test to observe the effects. Two Soviet EM-monitoring ships were even deployed during the Starfish Prime detonation with their own instrumentation to independently ascertain EM effects for Soviet purposes.

Of the 266 dedicated instrument monitoring stations, 156 were on land. There were 10 ships dedicated directly to data collection for the blast for the team located at the launch point on Johnston Island, with the ships carrying a total of 80 separate project-specific monitoring stations. 15 aircraft were deployed with the specific mission of monitoring the blast and collecting a wide array of data (primarily electromagnetic). The 15 aircraft carried a total of 30 monitoring stations aloft dedicated to different data collection protocols.

Besides collection stations on the immediately adjacent islands, there were additional monitoring stations on islands in the South Pacific and North Pacific. Instrumented ships launched from other islands to take advantage of the data collection opportunity. Three aircraft were launched from the island of Fiji (a KC-135 and two RC-121s) to gather data. Many towers were put into service on multiple islands to collect EM data for a large variety of instruments. A dozen of the instrumented test aircraft launched from Honolulu.

Operation Fishbowl (the umbrella program for Starfish Prime and other high-altitude nuclear detonations) aimed to study atmospheric ionization from nuclear bursts amongst other things. It was important to the U.S. government to ascertain what type of EM disruptions would occur from a high-altitude megaton-category blast that would affect U.S. offensive and defensive systems as well as civilian infrastructure. Most of the anticipated disruptions were related to EM effects, so EM studies were a large part, or the primary part, of the testing.

A large EM collection dish on Johnston Island mapped EM frequencies of operational importance during the blast. Aircraft and ground stations measured radar/ electromagnetic clutter. A search was done for EM noise that would affect the performance of radar and communications devices.

Reported EM Effects from Starfish Prime

There was a widespread scouring of military and civilian sources to ascertain EM and other effects after the Starfish Prime blast. Information was obtained from the vast amount of dedicated instrumentation and from queries to the public for anecdotes. Effects on civilian infrastructure were especially deemed crucial for a determination of what could be expected from enemy nuclear attacks.

Isolated Anecdotes in Response to Official Queries

With all the requests for input and the widespread review of all electrical systems because of the hype and search for damage, there were various anecdotes that found their way into different documents. None, however, was perceived to be an example of a mass-damage EMP effect other than the streetlight incident which is the main subject of this book. They were just random anecdotes of things not indicating widespread damage. Some of the things mentioned as having allegedly occurred during the Starfish Prime blast were: a rectifier being affected in a radio receiver, a retractable long-wire antenna on a test aircraft failing, communications being briefly attenuated or enhanced on certain frequencies because of enhanced absorption or reflection of EM waves in the ionosphere, radar reflections experiencing some minor path distortions or "jitter" when tracking missiles fired through the nuclear fireball, and a possible failure of a telephone microwave repeater link on the Hawaiian Island of Kauai during the peak of heavy phone call activity surrounding the blast.

The researchers were looking for anecdotes of failures. They received word-of-mouth commentaries in response to their queries, but nothing reported was considered to be verifiable or a large enough effect to be noteworthy, other than the Hawaiian street lights. Even this was seen as a small effect not worth studying until decades later. But, it was the best shot for proving nuclear EMP effects on electrical infrastructure, so it was eventually studied specifically in the Sandia report.

No Damage to Radio or TV Systems

One of the best indicators of a lack of wide area EM effects from any of the high-altitude tests was the official conclusion of the Los Alamos Lab in a 1976 study regarding any damage to civilian radio and TV systems. Radio and TV systems are sensitive devices designed to receive and amplify EM signals and can be overloaded and damaged. Los Alamos reported that, "There was no apparent increase in radio or television repairs subsequent to any of the Johnston Island detonations." [Quote from: "US High Altitude Test Experiences," a 1976 study of the Los Alamos Scientific Laboratory commissioned by the United States Energy Research and Development Administration.]

Civilian radio and TV systems are much more numerous than military systems and represent a much broader test-bed for EM propagation effects on communications and damage to sensitive electronics than any battery of tests deployed by the military. It is also a more impartial test-bed since the military and its contractors are, by their very nature, biased in their conclusions. They tend to desire outcomes that will foment a crisis that requires a cash influx to mitigate.

Civilian broadcast frequencies in 1962 also covered a very wide portion of the electromagnetic spectrum from 530 kilocycles to 1.7

MHz on the AM band, 2.3 MHz to 26.1 MHz on shortwave bands, 54 MHz (VHF) up to 890 MHz (UHF) on the TV bands, and 87 to 108 MHz on the FM band. Two-way radio frequencies dedicated to ham radio, business, "public service," and aircraft bands filled in the gaps in this list. That covers a lot of people using a lot of EM frequencies with a lot of sensitive equipment hooked to antennas sticking up in the air. The broadcast receivers and two-way radio equipment dedicated to all these frequency bands kept operating during and after the Starfish Prime burst.

Many speculate that damage to radio and TV systems would be from the alleged "E2" EMP that would be of an intermediate duration, similar to the length of a lightning strike. Some researchers have concluded that the E2 pulse is what damaged the strings of Hawaiian streetlights by saying that the wire lengths at the correct angles weren't long enough to receive an induced electrical flow from a longer "E3" pulse that allegedly damaged a long wire system in Kazakhstan during a high-altitude Soviet blast.

The non-existent damage to radio and TV systems in the Hawaii Islands and throughout the Pacific region would be a strike against a theory of mass EMP effects. But, of course, theories vary and some would say that the sensitive electronics in radios and TVs would be more affected by an alleged E1 pulse (which supposedly occurs alongside the E2 and E3 pulses from each blast). If so, they weren't affected by the alleged E1 pulse from any of the high-altitude blasts under Operation Dominic / Operation Fishbowl, including Starfish Prime.

No Mass Electrical Failures on Ships, Islands, or Aircraft

Besides the electrical devices and infrastructure on land, there were ships and aircraft operating at distances from the blast varying from directly underneath to hundreds and thousands of miles away. Ships and aircraft are packed with electronics. None of them were "bricked" and debilitated by any of the high-altitude blasts. No mass failures of electronics on ships, aircraft, or ground occurred immediately underneath or off to the side of the blast area at any distance.

Cyclotron Frequency Never Detected or Reported

Flashing back for a moment to the magnetobremsstrahlung section of this book; why wasn't the supposedly damaging cyclotron EM frequency ever detected by all the instrumentation deployed for the many high-altitude blasts (including the Soviet tests discussed later)? It was never detected or described despite the large number of EM monitoring stations sweeping many frequencies. If it is at a coherent frequency or frequencies, it should be verifiable if it is strong enough to cause mass damage. Things like oscilloscopes and frequency counters were in use during high-altitude testing. Instruments were scanning the whole EM spectrum and automatically recording their findings. Where was the huge burst of damaging radial EM radiation also known as cyclotron radiation? What was the frequency? If it was at a specific frequency (it must have been if it was producing a strong EM effect with coherent fluxing waves), what was that frequency? It should have been the strongest EM signal on Earth at that time. Why didn't it show up as the biggest spike on the scope? Tell me the frequency.

Who picked it up on a piece of monitoring equipment? Why didn't a lot of instruments pick out the frequency of the cyclotron emissions from other blasts, since these emissions would allegedly occur in all high-altitude blasts? Was it 10 megahertz? Was it 1000 megahertz? Was it 10 gigahertz? Was it 500 kilohertz? If it wasn't at a consistent coherent frequency, then it would be random and we would be back to self-cancelling random photon emissions. That's what they worked so hard to avoid with their fancy theories. These photons couldn't be random if they were going to induce an electrical force, so they jumped to magnetobremsstrahlung to try to get coherent waves at a low enough frequency to excite an electrical force in wires. What is that frequency? If it "varies" for each spiraling electron, if the electrons don't radiate fluxing EM energy in unison, then the theory is worthless, because the photons from each electron would arrive out-of-phase with all the other radial photon emissions spinning off all the other supposedly spiraling electrons being pulled downward along the Earth's magnetic field and they would be self-cancelling.

If the alleged cyclotron radiation produces multiple or non-specific frequencies for each photon that cannot be detected as frequencies of separate waves of photons, then we are back to self-cancelling photons or interleaving self-cancelling EM waves arriving at multiple intervals and out of phase with each other.

That frequency should not have remained hidden. It should have been captured on instruments that were sweeping the EM spectrum during these blasts. It should have been extensively discussed. The problem is that it wasn't there. There was no emission of coherent EM waves that could induce an electrical force in wires on Earth. It never happened. The fact that the cyclotron frequency is never detected, reported, or even discussed is not a big deal for any of the EMP crisis researchers. It is essential, however, if the theory is real. It matters to people like me that are being lied to. But, no one on the graft pipeline will go that far down the trail and pretend that this thing really works. What they have come up with theory–wise is more than adequate to please Congress and the high-ups in the military.

Swindlers Love Complexity

You can't, as a lying treacherous swindler, carry every ridiculous theory to its logical conclusion or you would expose your own fraud or idiocy. You have to stop the expression of the theory at the point of peak complexity. Bringing it back around to the real world and making it verifiable is self-defeating for hustlers. Forget the non-existent cyclotron frequency. Forget I even mentioned it.

23

Jokes

A very surprising portion of the official Sandia report was a section dedicated to joking descriptions of "EMP" contributed by the participating federal agency and contractor representatives at Sandia. These jokes belittled the idea of a nuclear EMP and admitted that is was a farce and a source of lucre for the government-connected Sandia folks. It was surprising to discover this content within a scientifically worded study replete with complex calculations aiming to *prove* the existence of EMPs. Here are some of the facetious, but insightful, comments made within the group of professional colleagues at Sandia that are actually *included in the text of the official report*:

> EMP EMPhasizes EMPire
> EMP EMPhasizes EMPloyment
> EMP is not real (is EMPty)
> EMP needs EMPathy
> EMP ... an Electro Magnetic Penguin

Those observations about what the letters "EMP" really signify to the government researchers are probably the most straight-forward and accurate statements in the whole Sandia report. We are at another point in this book at which I could say, "Case closed." Even the Sandia researchers in their own official report say that **EMP is not real** and that **EMP EMPhasizes EMPloyment.**

Wow! Spot on!

24

Are Modern Circuits
Susceptible to HEMPs?

Supposed Susceptibility of Smaller Modern
Electronic Components to EMP

This is an internet click-bait subject that we are bombarded with
nowadays. The never-explained theory amounts to a conclusion that
smaller modern solid-state electronics and integrated circuits must
be less robust than the 1940s / 1950s / 1960s era tube components
that existed during the nuclear blast heyday. The presumption is that
older tube-type electronics could withstand more HEMP thrashing
without failing despite numerous high-altitude nuclear bursts in that
era.

Since nuclear "EMPs" have apparently done nothing when interacting
with 1962 era electronic systems and with long-wire systems,
certainly we can change our focus and surmise that modern
electronics are more fragile and therefore more susceptible to EMP
damage allowing us to restore the EMP threat to its rightful status.

If you accept the notion that miniaturized circuits will be blown to
smithereens by an EMP, an expanded argument points out that
former mechanical or electro-mechanical switching of larger circuits
in the "power grid" have either been replaced by, or are controlled
by, these supposedly fragile non-tube circuits; the smaller circuits
turning the bigger circuits on and off via relays. These, the argument
goes, would certainly be more sensitive to EMPs than the older larger

tube circuits. This presumption then allows them to conclude that all power would be shut off and every type of consumer or industrial electronic or electrical device would fail. They would either fail from the zapping of their own internal micro-circuitry or from the failed "grid" that is controlled by micro-circuits. You see, they really need to scare civilians in their homes relying on the "power grid" that services them in order to get the fear train rolling on the nuclear EMP threat.

In other words, in the event that you notice that nuclear EMPs did not cause widespread failure of 1950s and 1960s era electrical or electronic systems owing to Starfish Prime or any of the other U.S. or Soviet high-altitude nuclear tests, they have concocted new unproven arguments about how there would be a threat now even if there wasn't back then.

Stoner Fodder

At this point in the chain of alleged events, an even more elaborate disaster is described when it is theorized that we wouldn't even be able to make new small circuits for up to a decade since factories that make small circuits are run by small circuits themselves (which would also have been destroyed). So, we would have to start from scratch to produce the technology by hand to make small circuits all over again that could operate factories to make more small circuits. We wouldn't have the electrical or electronic infrastructure to get going on the process, so it would take a long, long, time. Or, alternately, some doom-and-gloomers say we would have to start over again by importing the small circuits from other countries that weren't affected by the EMP—and that those countries may not want to sell to us because they are our enemies. These adversaries could then keep us living perpetually in the horse and buggy era.

It is a retrograde technology prediction made by fear mongers. It is the sort of paradoxical conundrum that stoned people like to propose and then waste time contemplating like, "Can you imagine how long it took ancient people to make the first straight-edge tool before they had straight-edge tools?" Great bar-room talk, but pointless, time-wasting, and meaningless. At about 3 AM during the EMP conversation, the stoner genius in the room can talk about an EMP-proof vault deep in the Earth that will have protected hermetically sealed sample circuits that future generations could retrieve as the seed material to bring us all back from the abyss.

Theory Pertains to Long Wires and Big Circuits Not Short Wires and Small Circuits

The main problem with all this is that conventional EMP theory has almost exclusively been argued and articulated as being a phenomenon that affects long wires. In practically every scientific iteration of the theory, long wires are supposedly needed to capture the electrical force from an EMP. Smaller circuit size has always been seen as something that makes electrical systems immune to EMP. The "science" or pseudo-science behind EMP relates to Faraday's Law of Electromagnetic Induction and therefore forecasts failures in long wire systems that could more readily receive an induced voltage / current, not failures in small circuits in small electronic devices. Most academic discussions of EMP effects even excuse incidents of non-damage because the involved wires "weren't long enough."

It is worth noting, however, that neither type of system—long wire or small electronic circuit—has ever been the subject of widespread EMP failures.

Problem for EMP Advocates: Modern Devices Have Shorter Internal Wires

It is assumed that large lengths of wire, even coiled wire like that in transformers and RF coils, would be more likely to induce currents from EM sources. This is, in fact, true. That is why there are long wire-wrapped coils in transformers. They are designed to receive induced EM currents from a primary EM circuit. But, this fact also presents a problem for those who want to see larger EMP effects in modern "smaller" electronics. Smaller devices also have shorter wires which are less capable of capturing currents from EM sources.

Modern Devices Operate at Higher Frequencies

Modern electronics operate at higher frequencies and therefore use smaller internal RF coils and smaller antennas; both of which present less metallic capture area that could receive an induced electrical force originating from an external EM source. Older electronics operated in lower frequency portions of the radio spectrum, which required longer coils of wire in internal RF coils and longer wires coiled around internal ferrite bar antennas or longer telescoping antennas or external antennas. Lower radio frequencies have longer wavelengths and require more wire in the circuit to function. Those old enough to remember will recall that old-style radio and TV devices were more likely to be affected by electrical and EM interference in the atmosphere. This is because of the lower operating frequencies and the tendency to pick up stray EM effects that would interfere with the intended signal.

Many people receive their entertainment, communications, and information services via internet signals provided by their cell phone carriers. These are in the higher frequency bands and require less inductive metal to be protruding into the sky or to be contained in the devices themselves. I can see no proof that systems like cell phones will receive an induced parasitic electrical current from a nuclear EMP. Even if a person accepts conventional EMP theory, the small modern devices don't fit the profile of a long wire system that would allegedly be impacted by a nuclear EM pulse

Cell phones operate in the 800 MHz, 900 MHz, and 1Ghz+ ranges. Therefore, they use very small internal wiring and coils. These frequencies are much higher than the common radio, TV, and communications frequencies of yesteryear. The larger circuits back then definitely made a bigger EM target. Over-the-air TV antennas are much less common than they used to be. Why? Because more people receive entertainment via satellite signals using smaller antennas operating at much higher frequencies.

Sirius XM radio uses EM signals in the 2.3 gigahertz range while DirecTV and Dish Network use 12 gigahertz signals. This can be contrasted with common communications and entertainment frequencies in the 1950s and 1960s that typically ranged from 530 kilohertz to 890 megahertz for two-way radios, broadcast radio, and TV. Those frequencies, and the devices that receive them, are still in use, but not to the extent they used to be.

The inductive coils and other electronic components are much smaller when operating at higher frequencies. They have less wire and less ability to receive induced currents from a "nuclear EMP" that allegedly requires long wires in the recipient circuit to cause damage.

There are still long wires out there in phone, cable TV, internet, and power distribution systems (as there were in 1962), but I can see no proof that a single pulse can, will, or ever did induce large voltages and currents in those long-wire systems that would cause them to fail over a large area.

Modern Electronics Use Less Power; Less Wire in Transformers

Old tube operated systems required lots of power and large power sources, so they came with embedded heavy transformers with hundreds or thousands of feet of wire wrapped in coils for the very purpose of capturing electrical forces from EM energy emanating from primary transformer windings. Conversely, modern electronics use very little power and have very small transformers. Smaller transformers mean less wire with a correspondingly lower tendency to be impacted by any external EM waves that would induce any unwanted electrical forces. Many voltage transformers are now digital resulting in much lighter TV sets and other electronics which now lack the long-wire internal windings needed previously.

Older Electronics Are More Likely to Capture External EM Radiation

Those large wire windings were present in 1962 electronics during Starfish Prime and were specifically designed to impart or receive induced voltage and current from fluxing EM energy. In my mind, those long-wire transformers that were present in most consumer electronics in 1962 would have more likely become parasitic inductive circuits to compatible fluxing EM pulses if such a thing had strongly presented itself in the adjacent local environment. It didn't. But, the circuitry was more susceptible to undesired EM effects back then.

In summary, more EM-inducing wire in older electronic devices owing to operation at lower frequencies with larger power requirements made them more susceptible to induction of unwanted currents and interference from external EM sources—like an alleged "EMP." Both the larger transformer windings and larger RF coils in older electronics worked to enhance EM induction. That is why you were more likely to "hear" and "see" atmospheric and other EM interference on your electronic device back then than you are now.

Solid State Electronics Were in Widespread Use during Starfish Prime Blast

The most significant fact related to the whole solid-state versus tube debate concerning EMP susceptibility is that solid-state radios were the predominant kind being sold by the late 1950s. By 1962, the year of the Starfish Prime blast, thousands of solid-state (non-tube) devices were in use across the Hawaiian Islands and other Pacific Islands. As mentioned previously, there were no increases in repairs of radios and TVs as a result of any of the Johnston Island tests— which included Starfish Prime. All of the high-altitude Johnston Island tests occurred after solid-state technology was commonplace. This should be the most essential point in the tube versus solid-state debate and should put to rest the claims that solid-state electronics would be affected by a nuclear EMP even though tube electronics were not.

YouTube Experimenters

Some YouTube users, wanting to join in on the EMP craze have tried to fry iPhones and other modern electronics with what they call an EMP. What they use are electromagnetic coils placed directly on the surface of the devices in an attempt to use EM photons to push electrons through the wiring in the devices per Faraday's principle. These experimenters often use a coherent fluxing EM field in their devices that is close enough to the device to possibly make a directional push in the internal circuits if per chance an internal circuit was aligned within the external field. The EM fields used in the videos, which are vastly different from scattering photons and electrons in the atmosphere originating from a cosmic source or a high-altitude nuclear blast, are exerting an EM force on the circuitry in the modern phone or electronic device from a nearby coherently-aligned and fluxing EM field. This is not what a nuclear "EMP" is. We already know that fluxing EM fields can induce an electromotive force in a nearby aligned circuit, so the experiments are fairly pointless.

But, even when the on-line videos make apples-and-oranges comparisons and attempt to show EMP effects on consumer electronic devices, like iPhones, they rarely inflict any lasting effects.

Modern EM Shielding

I have seen YouTube users, surprised at the lack of lasting effect, disassemble the modern devices and discover, to their surprise, metal shielding plates that are often even labeled as "EM" or "EMI" (electromagnetic interference) shielding. They take apart other devices with modern circuits like computers and discover metal plate EM shielding around circuits in those devices as well. This has been done for years. It is a version of a Faraday Cage that blocks EM energy. It was done in the past, but is done even more thoroughly in the modern era. This shielding is designed to prevent EM interference and unwanted EM energy from going into and radiating out of electronic circuits. It was harder to shield whole larger sprawling circuits in the past. Various sensitive components spilled across multiple large circuit boards with large interconnecting wiring. All of those sprawling components and wiring presented natural EM targets. Only certain components were shielded, not all their peripheral semiconductor components and large interconnecting circuit traces and wiring. Nowadays, with smaller integrated circuits, whole circuits can be easily encompassed within an EM shield and be protected from EM energy on all sides.

Why are we to conclude that technology has gone backwards? Why must we believe that modern circuits are more prone to EM interference than they were in 1962? As a matter of fact, some things, as mentioned above, have changed over the years that make proposed nuclear EMP damage even less likely in modern circuits.

145

25

EMP Experts:
Nuclear Devices Are
Too Small or Too Big

Nuclear EMP: Kiloton versus Megaton

There are excuses made for the lack of widespread EM damage from both larger and smaller blasts. They are often articulated around the duration of alleged E1, E2, and E3 pulses. For example:

> The first two of the Soviet "K-Project" high-altitude nuclear tests over Kazakhstan in 1961 were only 1.2 kilotons (at 150 and 300 kilometers altitude), so the EMP could be carefully measured, but these tests, apparently, did not have much of an impact on the 1961 infrastructure of Kazakhstan. This is unsurprising because of the hardier electronics of that era (which would be less susceptible to E1), as well as the smaller E3 pulse from such small devices. [From "Soviet Test 184: The 1962 Soviet Nuclear EMP Tests over Kazakhstan" by Jerry Emanuelson.]

Once again, we see an admission that there was no impact on the infrastructure of Kazakhstan from the K-Project high-altitude blasts, along with a presumption (with no proof) that similar sized high-altitude blasts would affect modern electronics.

If smaller kiloton range high-altitude blasts didn't do the trick, researchers could at least focus in and look really hard for an effect from larger Soviet blasts in the same region. When EMP researchers did so, they still found no widespread effects, but they claim to have found a Hawaiian type unverifiable small incident. Also, an oft-repeated excuse and pet theory was trotted out explaining why large blasts defeat their own EMPs because of the two-stage detonation in thermonuclear devices causing "pre-ionization" (more malarkey to excuse the lack of widespread EMP damage from large nuclear devices while keeping the theory alive).

Emanuelson explains below (ibid.) why large blasts defeat their own EMPs after describing previously how the smaller initial K Project detonations weren't large enough to create damaging EMPs on the "hardier" electronics of 1961:

> It is clear from the data that has been released on the E1 component of the pulse that the thermonuclear weapon used in Test 184 was particularly inefficient in producing EMP. In all thermonuclear weapons, pre-ionization of the upper atmosphere from the gamma radiation of the first stage of the weapon limits the peak electric field generated by the final burst of energy; and it appears that the peak electric field produced by Test 184 was not much more than 10 kilovolts per meter over any point in Kazakhstan. If the weapon had been a simple single-stage pure fission weapon of the same yield, the fast E1 component of the pulse would have been 3 to 5 times the intensity.

So, the high-altitude nuclear blasts are either too small or too big to produce a damaging EMP.

To be consistent and for there to be any chance (per their theory) that a damaging nuclear EMP event can occur, it must be from a high yield weapon (20 to 100 times the size of a Hiroshima bomb), but not thermonuclear, i.e. not a more efficient modern H-bomb that can more practically produce those yields. So, just to test this EMP mass-destruction theory, which would also fail, a super massive old-style fission bomb—way bigger than any built before—would have to be built and sent into outer space on a very robust expensive launch vehicle with a total project cost of what—half a trillion dollars? How is this something that people fear from an impoverished "rogue state" that lacks the economic activity and tax base to support such a venture?

Not to mention that there have never been any EMP-produced mass system failures from any size nuclear blast, even after numerous detonations occurred over Kazakhstan—which was not a barren wasteland free of electrical infrastructure. But, theorists are not consistent on the view that bigness minus thermonuclear equals EMP. It is often stated nowadays that a *very small* nuclear device could knock out the electrical infrastructure for the entire continental U.S. due to "fragile modern electronics." So, take your pick. They say that big and little won't work for various reasons and they also say that big and little will produce devastating EMPs for various reasons.

No Results: Maybe Their Theory is Wrong

They think the effect must be there even though they never see it and nuclear devices of all sizes detonated at all altitudes never produce it. They gloss over the coherent flux needed for EM electrical induction and the inability of an "EMP" to present a different EM angle to all aspects of a recipient circuit and just conclude that coherent fluxing and anything else that some purist might want is present. Nobody is demanding proof, so keeping the unassailable theories alive is a pretty easy gig.

I personally think that maybe they should schedule a conference, dial up their pet nuclear Ph.D.'s, pull up their chairs, and have a round table chat to consider that maybe the EMPeror has no clothes since nobody ever sees them. No, wait a minute. That wouldn't work. Who would pay for the airline tickets, the salaries, the hotel bills, and the food for such a conference? Never mind.

26

Soviet EMPs?

Soviet K Project Tests

Soviet blasts like the high-altitude "Test 184" under the "K Project" were also studied extensively for EMP effects. Much has been made of the EMP effects, but when a closer examination is made, there is little to it.

These K Project detonations were conducted in good testbeds for EMP effects because, as Emanuelson points out (ibid.):

> These EMP-producing tests were done over a large populated land mass in Kazakhstan. Even though the economic state of Kazakhstan in 1962 was quite primitive by today's standards, it was heavily industrialized and electrified.

Emanuelson studied Test 184 (also known as Test K-3) specifically because, of all the Soviet tests, it was the one that theoretically was most likely to produce mass damage from EMP effects in that it was detonated at very high-altitude (290 km) and involved a large yield (300 kilotons).

Hundreds of Soviet Tests over Kazakhstan

There were hundreds of Soviet nuclear detonations in Kazakhstan from 1949 through 1989. There are seven that were of sufficiently high-altitude that they would have theoretically produced widespread damaging EMP effects, if such things were real. The Soviet tests are important in the EMP discussion because they occurred entirely over land. The alleged EMP mass-damage radius included areas with large populations and areas with significant civilian, military, and industrial infrastructure. Test #184 alone had a calculated EMP mass-damage radius covering the entire territory of Kazakhstan over which it was detonated. A list of Soviet blasts that were of sufficiently high-altitude (above 12 km) and sufficient power to theoretically produce (per conventional wisdom) powerful damaging EMPs is repeated below.

Soviet Nuclear Blasts That Allegedly Should Have Produced Widespread EMP Damage

USSR—Test #88; 10.5 kiloton detonation in 1961 at 22.7 km altitude

USSR—Test #115; 40 kiloton detonation in 1961 at 41.3 km altitude

USSR—Test #127; 1.2 kiloton detonation in 1961 at 150 km altitude

USSR—Test #128; 1.2 kiloton detonation in 1961 at 300 km altitude

USSR—Test #184; 300 kiloton detonation in 1962 at 290 km altitude

USSR—Test #187; 300 kiloton detonation in 1962 at 150 km altitude

USSR—Test #195; 300 kiloton detonation in 1962 at 59 km altitude

Stalingrad - No EMP Damage

The first two on the chart (Tests 88 and 115) were detonated 20 miles southwest of Volgograd (formerly Stalingrad) and should have wiped out the electrical infrastructure of the entire city when considering the blast radius calculations and damage estimates relied upon by the EMP crisis community—but, no EMP damage was reported. Even considering Soviet era secrecy, we would have certainly heard by now if an entire major city had been sent back to the Stone Age by an EMP. These two devices aren't the smallest or the biggest, but they are in a size category that modern EMP theorists say is more than big enough to wipe out the electrical infrastructure for the entire continental U.S. And yet, Volgograd was 20 miles away and—nothing.

Soviets Looked Hard for EMP Effects

The final three detonations in 1962 under the Soviet K Project used high power 300-kiloton warheads.

Emanuelson points out (ibid.) that:

> Like the U.S. Starfish Prime test and others, the 1962 Soviet high-altitude tests were monitored by a very large array of scientific instruments ... the EMP phenomenon was a major reason for the project. A number of instrument packages were launched during the K-3 and K-4 tests (Tests 184 and 187) using the Soviet MR-12 meteorological rockets. The small MR-12 rockets were timed to reach their apogee (highest point) of 130 to 140 km. at the moment the nuclear weapon was detonated.
>
> During all of the K Project tests, several rockets of different types with scientific instrumentation packages were launched within minutes of each other from Kapustin Yar, the Baikonur Cosmodrome, and from the Saryshagan test range.

The Soviet high-altitude tests were also monitored by both the Soviet Cosmos XI satellite and the United States Explorer XV satellite.

The official monitoring was, of course, in addition to the observations of any effects by the vastly larger unwitting testbed consisting of the entire population of Kazakhstan interacting with the whole electrical and electronic infrastructure of the region.

Emanuelson mentions (ibid.) when discussing the powerful Test #184 that,

> At an altitude of 290 kilometers above the detonation point in central Kazakhstan, the distance to the

horizon would have been more than 1900 kilometers, which would have caused an electromagnetic pulse that covered all of Kazakhstan, with strongest effects near the south central region of Kazakhstan. The world's first spaceport, the Baikonur Cosmodrome, is about 300 kilometers (190 miles) to the southwest of the detonation point, and with the orientation of the geomagnetic field over Kazakhstan, **the Baikonur Cosmodrome should have received some of the worst of the EMP effects, although nothing about this has been openly reported**. [Bold emphasis added.]

Once again, there was no observed or reported mass-damage EMP effect in areas that presumably would have had a concentrated effect based on calculations of the orientation of the Earth's magnetic field. There was no reported effect in the area of magnetic concentration in Kazakhstan, even when researchers wanted it to exist and theorized that it would exist.

All the Proof is Classified?

It is convenient to conclude (and EMP researchers often do) that the serious evidence proving mass-damage EMP effects is all either "classified" by the U.S. government, or retained in secrecy or obscured by government inefficiency or buffoonery in the former Soviet Union or by some other government entity that holds the key somewhere within masses of data, but doesn't know how to dredge it up and interpret it; or, that they are keeping it from us lest we tremble in fear at its revelation. This is sheer nonsense.

Mass electrical failures of the power and communications systems for the entire Kazakhstan region (or the Hawaiian Islands, etc.) would have been a spectacular event and would be discussed ad nauseam by EMP researchers, not to mention human rights advocates and anti-nuke protesters. And, it is not like the government to keep some big juicy money-generating crisis from the public because we might open our wallets and clamor for government solutions if we were to know. They are all about the crisis. Don't worry. They wouldn't hold it back if they had any way to scare you a little more and to prove EMP was real.

Soviets Reminisce, But No Mass EMP Damage

Emanuelson even gives anecdotes from the day-to-day lives of Russian soldiers and their families going about their business in the nuclear blast areas *at the time of the blasts*. Certainly mass society-wide extinction of electrical systems across Kazakhstan would have come out by now. The Soviet Union has broken up. People are talking, people are reminiscing, but no one remembers the mass electrical failures? That's because there were no mass electrical failures. We don't have to have access to a secret government "X-File" in order to detect mass failures of electrical infrastructure. Forget the pet scientific calculations for a moment which conclude that it should happen and observe that it *does not* happen and *did not* happen. There have been 2,476 nuclear detonations on Earth and there have been no mass-damage EMP events. That should be the end of the story. I should be able to say once again, "case closed."

Loborev: What Happened in Kazakhstan

Little was ever mentioned about EMP damage from Test 184 (or the other Soviet tests) until 1994 when Vladimir Loborev spoke to the EUROEM (European International Symposium on Electromagnetic Environments) Conference in 1994. In that presentation, he mentioned that a diesel generator and a 1000-kilometer underground power line it was attached to failed after Test 184. Although Loborev described a diesel generator, the reference is often upgraded by the EMP crisis community to "a power plant." The alleged generator failure also happened days after the blast, not immediately, so U.S. researchers speculated that delayed generator problems after a detonation would occur from a "dielectric breakdown" in the generator windings caused by the EMP. There was no generator to analyze, but they wanted EMP to be the cause so they made-up a delayed effect EMP theory.

This was a real stretch. There was not a mass infrastructure failing across Kazakhstan although calculations indicated there would be. The only purported failure attributed to Test 184 that was considered really significant for bringing EMP onto the front burner as a major influence on electrical infrastructure was the purported failure of a 1000-km (621 mile) buried line that was connected to the generator that failed (days after the blast). Fires at the location of the generator and the connected wire are mentioned by the EMP community, although this appears to be an interpolation because Loborev only mentioned overheating and a short circuit as the cause for the generator failure. The generator shorted out days after the blast and no longer provided power to the buried line. This has been exaggerated and repeated over and over to obtain the maximum effect from what was a non-event.

With regards to speculations that there may have been a fire or fires related to the line the generator was purportedly connected to, Emanuelson notes (ibid.):

> Most details about this underground line are very sketchy, and the reported length seems to be impossibly long for a single length of line carrying any kind of alternating current without some sort of re-generating station.

So, Loborev's reporting of the incident seems flawed and exaggerated from the outset, even though it described nothing near a mass failure of systems across the whole region. Loborev said in his non-native tongue to the conference attendees (ibid.):

> The matter of this phenomenon is that the electrical puncture occurs at the weak point of a system. Next, the heat puncture is developed at that point, under the action of the power voltage; as a result, the electrical power source is put out of action very often.

In other words, Loborev, trying to sound technical, says that electrical systems sometimes fail and that heat is observed during the failure which results in the system going down. Loborev seems clearly to be describing, in broken English, how a short circuit produces heat and can result in failure; a very common non-EMP occurrence in any electrical system. The "heat" mentioned by

Loborev is turned into "fires" by some in the U.S. EMP crisis community who want to document a bigger effect.

Conveniently, Metatech Corporation (an EMP-advocating U.S. company that does federal contractor work in the nuclear field for the U.S. Oakridge National Laboratory), was present and created enhanced notes with Metatech-added notations supposedly derived from Loborev's presentation. Metatech saw more clearly than Loborev that EMP sensationalism could be a dollar-generating phenomenon. As noted by Emanuelson (ibid.):

> Many of the additions to Figure 2 in Loborev's paper, as shown in the illustration below, are based upon notes made by Dr. William Radasky of Metatech Corporation at the time of Vladimir Loborev's original oral presentation at the 1994 EUROEM Conference. It is important to note that the illustration just below is a [sic] illustrative representation, and not a map. It does not accurately show the geographical directions to the indicated damage.

It is presumed by most EMP advocates that the alleged longer duration E3 component of the alleged nuclear EMP generated by Soviet Test 184 caused the alleged damage to the underground line and the attached generator. (Yes, I know I used the word "alleged" three times in that sentence, but that's the only way I can indicate that the dots they are connecting probably don't even exist.)

There were other random anecdotes of failures from Test 184 including some in radios and a phone line, but hard data is missing as evidenced by Emanuelson's comment (ibid.) when he tried to assess line length and orientation when reviewing the damaged phone line claim: "There are approximations in the published information that make it impossible to know the exact location of the far end of the line."

E1, E2, or E3: Take Your Pick to Explain Electrical Failures—or Lack Thereof

Emanuelson attributes the "failures" from Test 184 to the longer alleged E3 pulse and concludes that the damage in Hawaii was from a shorter duration E1 pulse emanating from Starfish Prime. His explanation involves the wire lengths bringing him to conclude that the shorter wires for the Hawaiian street lights (several blocks long) were affected by a shorter E1 pulse and the impossibly long 1000 kilometer wire in Kazakstan was affected by a longer E3 pulse. Emanuelson then forgives the longer lines in the power grid in Hawaii (many miles long) for not failing from a Starfish EMP by explaining that they were too short to be affected by an E3 pulse. Emanuelson (ibid.): "The E3 component of the Starfish Prime EMP had no reported effect on the Hawaiian power grid because the lines were too short."

So, per Emanuelson, the long power lines in Hawaii weren't affected because they were too short to be affected by an E3 pulse and, I guess, too long to be affected by an E1 or E2?

27

Los Alamos Study

Los Alamos: Starfish Effects Were Insignificant

Although the Los Alamos Scientific Laboratory concluded (as everyone else did) that an "(EMP) caused brief outages of a street lighting system in Oahu" [in a study titled "United States high Altitude Test Experiences: A Review Emphasizing the Impact on the Environment;" commissioned by the United States Energy Research and Development Administration; published June 1976], they also concluded that the effects of Starfish Prime were, in essence, inconsequential, saying that, "In summary, the effects of the US high-altitude explosions on the normal activities of the populations were either insignificant or under protective control involving little harassment or irritation."

So, Los Alamos, in that study, concluded that any effects on unprotected civilian systems from Starfish Prime were insignificant.

Los Alamos Lacks Confidence in Theoretical Treatment of Hawaiian Effects

The 1976 Los Alamos report (ibid.) confirms what the 1985 Mattox study and the 1989 Sandia study also concluded about Starfish Prime in their writings in the following decade:

> It would be desirable to present a still better, fully coherent story of the whole pertinent phenomenology. While today's knowledge of the late

phenomenology could conceivably be improved by putting more bits and pieces together—a tedious task—**the full picture would probably not evolve**, simply because of limitations in observational data. Furthermore, the **theoretical treatment of these late phases of the phenomenology and of the atmospheric interactions is difficult to do with confidence**. [Bold emphasis added.]

So Los Alamos, a decade closer to the facts, still concluded that there was not enough "observational data" to conduct a "theoretical treatment" "with confidence" of any EM effects coming from Starfish Prime.

Anecdotal Starfish Outages

As mentioned previously, Los Alamos pointed out that there was no increase in radio or TV repairs from the high-altitude detonations launched over the Pacific from Johnston Island. Los Alamos also mentioned in the same study that "No failure was noted in the telemetry systems used for data transmission on board the many instrumentation rockets." So, the data radios did not fail in the many rockets fired directly through the nuclear fireballs for the high-altitude tests.

Solid State Susceptibility?

Los Alamos chimes in with Emanuelson and the rest of the parroting EMP community about the susceptibility of modern electronics to EMPs. Los Alamos (ibid):

> With the increase of solidstate circuitry over the vacuum-tube technology of 1962, the susceptibility of electronic equipment will be higher, and the probability of more problems for future detonations will be greater.

Los Alamos and Emanuelson both conclude that there will be failures in the solid-state era that weren't present in the tube electronics era of the 1950s and early 1960s. This comes, after both of them conclude that failures would theoretically relate to long wires, not individual electronic circuits—which, by the way, have never had mass failures. It is a fallback position blaming fairies and goblins with no theory to back it up—in case the long-wire boogey-man theory doesn't work out.

Solid State Circuitry Was Widespread by Late 1950s

Might I point out that the data rockets fired through the nuclear fireball and the monitoring instruments used by the U.S. government to monitor the blasts had solid-state electronics in them? Actually, non-tube solid-state radios had become the mainstay of the industry by 1958, four years before the Starfish Prime blast. Why wasn't this large array of non-tube radio receivers which was operating across the Hawaiian Islands and in many other Pacific Islands demolished by an EMP from the Starfish Prime blast and the other high-altitude blasts in the Pacific?

28

Z Machine and Trestle Machine

The Sandia Labs "Z Machine" and "Trestle Machine" and other such machines around the world which were produced usually to make electrical pulses for nuclear research programs, are sometimes mentioned as devices that prove the existence and efficacy of high-altitude nuclear EMPs and their ability to inflict damage on electronic systems. These devices use completely different principles as those used to explain supposed nuclear EMPs.

These machines use capacitors to store large amounts of electrical energy and then release it through wires and grids acting as antennas and ground planes oriented around a focal area. This focused close-proximity instantaneous release of millions of watts of stored electrical power on a target (like a vehicle) at appropriate angles is not at all what a high-altitude nuclear EMP is said to be. The gamma rays supposedly resulting in Compton Scattering and radial electromagnetic radiation from magnetic braking at high altitude are, of course, absent and are not the basis for the functionality of these machines.

We already know that direct electrical discharges like lightning can result in damaging electrical flows when they come into contact with circuits. We also know that fluxing EM waves can produce electrical flows in nearby aligned circuits. We are not talking about either of those situations with nuclear "EMPs." An up-close high-power focused electrical discharge from something like the Trestle Machine or the Z Machine does not simulate a cascade of events from the high-altitude release of gamma radiation stripping electrons off atoms in the atmosphere that then flow along the magnetosphere of the Earth producing radial radiation that can arrive at a circuit from a long distance and find all the correct angles to push electrons in a circle through that recipient circuit. That is a fantasy that is not proven by constructing a device consisting of large capacitors and generators hooked to wire grids designed to release massive amounts of stored electricity upon nearby electronic devices.

Pop culture TV documentaries, like an episode of "Future Weapons" on the Discovery Channel, that show supposed EMP simulating machines (in that case the "Horizontally Polarized Dipole II") disabling vehicles, make similar inappropriate comparisons to scare viewers. The technology does not even come close to re-creating the theoretical chain of events that allegedly produces a high-altitude nuclear EMP that would supposedly knock out the power grid from sea to shining sea. These "EMP Machine" scenarios make apples and oranges comparisons that we are supposed to see as validation and proof of nuclear EMP effects.

29

Summary

In this section, I will revisit and summarize the arguments made in this book against the existence of a mass-damage nuclear EMP phenomenon in general, and in the Hawaiian Street Light Incident in particular.

Gamma Radiation Doesn't Induce an Electrical Current in Wires

The EM photons of concern (to EMP advocates) emanating from a nuclear burst are said to be in the form of gamma waves. The frequency of gamma waves is too high to excite a current in a metallic conductor like a wire on Earth since the plasma frequency in the metal is below the frequency of the gamma radiation. This fact discredits the idea of an EMP originating from a high altitude nuclear blast.

Scattering Photons Do Not Induce Current

If, per EMP theory, photons in gamma radiation are said to be acting upon the atmosphere as individual colliding and scattering particles, their arrival times and impacts on atoms in a wire on Earth are random, self-cancelling, and don't fit Faraday's Law of Electromagnetic Induction. Any electrons ejected by photon impacts in a wire would go in random directions. They would not produce an electrical flow in wires. This other aspect of photons—acting as particles as opposed to waves—can therefore not produce an EMP either.

EM Waves from One Side Would Not Cause Directional Electrical Flow in a Wire

If coherent EM waves really emanated from a nuclear blast and then swept a wire loop, the effects on electron flow in the circuit would be inconsequential and non-directional since the EM waves arriving from outside the whole circuit sweep the back side of the circuit in the opposite direction at the same time they sweep the front side. This would create an opposing force that would inhibit electron flow. All electrical systems are in circuits (loops). So, an EM effect would have to present the appropriate angle to all sides of the loop in order to push electrons through it. This is a major problem for EMP advocates that describe an EM force arriving from one direction outside the circuit.

A Pulse Is Not a Wave

The word "pulse" is used by EMP researchers to imply coherent EM "waves" needed for electrical induction in wires. Yet, there are no demonstrable EM waves from a nuclear blast of a type that can induce electromotive forces in wires on Earth.

E1, E2, and E3 Are Used As Excuses, Rarely as Explanations

When E1, E2, and E3 are actually used in official studies, they are usually used to excuse non-effects, e.g. "this wire was too short to be impacted by E3 or too long to be impacted by E1." They are often used to postulate unobserved effects, but rarely used to discuss recognized widespread historical effects in the real world—since such effects are non-existent.

Tiny Alleged Failure in Light System Represents the Biggest "Proof" of EMP

It is clear that only a one percent failure of a certain subordinate type of street lighting in Honolulu was the extent of the claimed systemic failure of the civilian electrical infrastructure in Hawaii. That is clearly not indicative of a mass damage effect.

Other Long-Wire Systems Had No Failures

Other long-wire systems in Honolulu should have had failures if EMP theories are legitimate. Conventional EMP theory is based on the idea that long wires will receive induced electrical forces from an EMP. The fact that there weren't failures in the other pole mounted long-wire systems in Honolulu pertaining to things like telephone, cable TV, power, mercury vapor lighting, police call boxes, fire alarms, etc. is an indication that there was no EMP effect at all from Starfish Prime.

Other Fragile Systems Would Fail Before Street Lights

Other fragile devices connected to the power grid did not fail. Street lights are more robust and tolerant of voltage variations than many other devices like communications equipment. The fact that those other systems connected to the power distribution system did not fail is an indicator that there was no EMP effect.

No Records

Another serious issue is the lack of records. They had all been destroyed. Sandia admitted that they had no records to study and yet proceeded to make guesses about the Honolulu system that would be favorable to an EMP cause. Sandia used what it labelled as "imaginary" data to make its conclusions. This fact makes the study entirely unscientific and renders any conclusions invalid.

Affected Oahu Street Light System Non-Existent at Time of Study in 1980s

The street lighting system in Oahu had been totally changed by the time researchers begin to study it in the 1980s. There were no records to review and no actual systems to study. This prevented any conclusions from being drawn from on-site studies.

Failed On-Site Survey: Azimuth Not Ascertainable

Azimuth angle of the wires in relation to the blast was considered to be crucial to the researchers in the official Sandia study. Yet, azimuth angle could not be determined since the 1962 system in Honolulu no longer existed at the time of the study. Mattox attempted to study a representative string of lights in 1985, but was unable to do so since the system had totally changed. This should have been the end of the story for the researchers. They should have concluded at that point that an EMP cause could not be determined since the "essential" azimuth angle they wanted could not be determined. And yet, Sandia ignored the facts from this only on-site visit and concluded that the correct azimuth must have existed.

Pre-Ordained Outcome

The Sandia study had a pre-ordained outcome. A strong bias by researchers is a big reason to discard the conclusions of any study. The researchers themselves would get more money for their area of endeavor if they could prove a nuclear EMP. We should be suspicious of the conclusions of crony industry puppet-masters who want control of the taxpayer money spigot. The federal contractors paid to conduct the study wanted to preserve the fear status of nuclear EMPs. They were concerned that it might "lose stature as a threat."

Multiple EM Monitoring Stations Did Not Find a Mass Damage Effect

The Starfish Prime blast was the most thoroughly monitored of any nuclear blast in history and was the most powerful nuclear blast in outer space ever attempted by mankind. It used the quintessential recipe for EMP production per all the theories—and the best of ingredients: high-altitude and high power. Everything was there. The sky went up in flames from the perspective of the onlookers in Hawaii. It is hard to imagine a better test of any possible EMP effect on civilian infrastructure; one that would hardly be tolerated in our day and age. Yet, there was no EMP mass damage effect detected. Unverifiable 1% damage in a single subordinate system in Honolulu doesn't count as a wide area mass effect. If they couldn't get a damaging EMP from Starfish Prime, the most heavily instrumented and mother of all high-altitude, high-yield blasts, then such an effect is clearly bogus.

Poorly Run Lighting System Was Cause of Failures – Not EMP

The poster child for EMP damage to civilian infrastructure is the "Hawaiian Street Light Incident." Yet, this tiny occurrence happened within a shoddy third-world style hodge-podge system that was literally held together with popsicle sticks. This system had to fail, and did fail often, as evidenced by the comments of Sandia researchers regarding the continual fuse modifications and repairs. Sometimes the fuses even failed from excess voltage upon initial insertion until more layers of plastic were added within the fuse to make a larger gap. The system was failure prone. Any small number of failures that "occurred" on a night when everyone was looking for failures and trying to garner hero status as a talebearer is insignificant. The utility company employee that was interviewed by Sandia indicated that employees didn't always report their repairs. So, there is no baseline for what is a normal failure rate.

This poorly designed and run lighting system was clearly the probable cause of any small number of failures that occurred on any particular date. Some of the things identified by Sandia as being problematic in the Honolulu street light system which would lead to non-EMP failures are 1) Varying high voltages within same system, 2) Popsicle sticks routinely used to hold contacts open on high voltage transformers, 3) Voltage variance in each string due to varying number of lights, 4) Confusing mix of voltage transformers used to deal with varying source voltage and varying load voltage from varying bulb counts in each string of lights, 5) Varying lumen rating for bulbs which affected the current draw for the different bulbs and for the entire string, 6) Lack of desire within company to purchase proper current regulators, 7) Regular informal hand-modifications to fuses using files and stackable plastic washers which masked inappropriate line voltages on each string of lights, 8) Variety of fuse sizes purchased from a variety of manufacturers, and 9) Lack of consistent guidance to employees and lack of standards as evidenced by inconsistent information on fuse burn-through voltages seen in company literature.

Series-Wired Circuits Are Prone to Failure

The entire string of lights on a series lighting circuit is extinguished when an individual bulb in that string burns out. Each bulb is an essential part of the circuit. If the filament in that bulb burns out, the circuit is broken and the whole string goes dark. Sandia conveniently obtained an unverifiable anecdote that none of the tens of thousands of bulbs were burned out on the island of Oahu that night. This precluded the argument that the small number of string failures was due to individual burned out bulbs —a normal occurrence.

Ancient Anecdotes Were Key for Sandia

Sandia had no records to review so they conveniently relied on modern anecdotes drawn out of a former employee at the time of the study regarding his memories from 27 years previous. Those single witness anecdotes conveniently gave Sandia whatever facts they desired. Anecdotes which involved impossibly broad observations and conclusions on the part of the former employee became key for Sandia to get the result it wanted. This type of anecdotal "evidence" —without any physical evidence—is not scientifically conclusive whether you are talking about abduction by space aliens or EMPs.

"Coupling Analysis" Section: Garbage In, Garbage Out

The "Coupling Analysis" section of the Sandia study involved calculations that used invented data to obtain a desired result. As the saying goes, "Garbage in, garbage out." If the calculations were populated with imaginary data, then none of the conclusions from "coupling analysis" of EM effects with wires are meaningful. Sandia even concluded that, "Detailed calculations are not justified because of such uncertainties." But, that didn't stop them. They went ahead and made up numbers for their detailed calculations. That one fact should void the whole study and any conclusions it contains.

Sandia Director Rejected EMP Theory

Sandia Director of Testing Don Shuster rejected the claims that a nuclear EMP caused the failure of a small number of street light strings in Honolulu. He offered a logical theory within the official

Sandia report that was at least as plausible as the one that the team arrived at, probably more so. His theory describing failures from extra start-up demand on the system from simultaneous photocell activation of street lights after the daylight condition created by the nighttime blast was deserving of more than the quick dismissal it received.

"Everyone Awake" Theory is More Plausible than EMP

Another plausible non-EMP theory relates to the fact that most people who were normally asleep in Hawaii were awake to see what would happen from the large nuclear blast in the sky. They were using power during their waking activities and this power coupled with the power used by normal night-time usages, like streetlights, may have stressed the system and caused the small number of failures that were observed. Excess demand is a common cause for electrical failures. It seems eminently more likely as a cause than a fanciful never before seen leprechaun.

No Damage to Radio or TV Systems

The civilian population in the Pacific was the biggest testbed for any EM effects coming from Starfish Prime and other Johnston Island high-altitude nuclear blasts. Per the Los Alamos study, there was no noted damage, however, to civilian radio and TV systems. These are sensitive devices that are designed to receive and amplify EM signals. They can easily be overloaded and damaged. They would have certainly been affected if this monstrous EM force was present that was supposed to knock out electrical devices across a large swath of territory consisting of millions of square miles. The same can be said for ships and aircraft. They survived the blasts as well with their electrical systems intact.

Los Alamos: Starfish Effects Were Insignificant

Los Alamos accurately described the scale of any effects from Starfish Prime on the Hawaiian Islands as insignificant. Since Los Alamos built the bomb, they probably didn't see any benefit in blaming themselves for any alleged damage. Sandia, however, dwelt little on the subject of the magnitude of any failures—which was so small as to be routine and insignificant. Instead, they dove into mathematical calculations using made-up data to get their EMP result.

Los Alamos Lacks Confidence in Theoretical Treatment

The 1976 Los Alamos report also concluded that a theoretical treatment of any failures in Hawaii would be difficult to do with confidence because of the lack of data. This was a decade before Mattox and Sandia tried to look at the situation. There was even less data to look at in the 1980s. Los Alamos' assessment of the situation should be given more credence. There was nothing that could be studied with confidence. Sandia's conclusions should have emphasized that fact.

Other Hawaiian Islands Not Affected

Why weren't the electrical systems in the cities on the other Hawaiian Islands affected? Other Pacific Islands had no electrical infrastructure damage either. This is a clear indicator that the Hawaiian Street Light Incident involved a normal everyday small failure that was not caused by a nuclear EMP.

Modern Electronics Less Likely to Capture Induced EM voltage

Since long wires have not had the damage that was postulated, some have begun to put their hopes in small electronics as the new potential victim of EMP. They say that the larger 1940s / 1950s / 1960s era tube electronics were more robust than our modern electronics which would be fried. The scientific theories have not been modified significantly, however, and still pertain to long wires. That being said, there is no proof that small modern electronics would be affected. Modern electronics are actually better at rejecting stray EM forces than their older counterparts. Modern electronics have shorter internal and external wires because they operate at higher frequencies that use smaller RF coils and smaller antennas. They also have less internal wire since they use less power, which obviates the need for large transformer windings which by their very nature are designed to receive EM currents from induction.

Less wire means less EM electrical inductance and more immunity to EMPs, if such things were real. Modern electronics also have extensive EM shielding which has improved over the years. Even amateur YouTube experimenters have been surprised at the lack of lasting effects on modern electronics when they try to damage them with homemade "EMPs." Researchers often say that solid-state electronics are more susceptible to nuclear EMPs than the "tube" systems of the 1940s, 1950s, and 1960s. This whole objection is nullified by the fact that solid-state electronics were in widespread use starting in the late 1950s. Most radios being sold during the previous four years leading up to the Starfish Prime blast were solid-state. They were not affected by the blast.

Soviet Tests Should Have Wreaked Havoc

If EMP theory is true, the high-altitude Soviet tests over Kazakhstan should have devastated the extensive electrical infrastructure lying beneath them. Yet, they happened repeatedly without that result. There were hundreds of nuclear blasts over Kazakhstan. This is another proof of the non-existence of society-wrecking EMPs. The Soviets looked hard for wide area EMP effects also. There simply weren't any.

Thousands of Nuclear Blasts Never Produced Mass Damage

The Starfish Prime detonation was part of a series of high-altitude nuclear blasts that augmented other nuclear testing occurring at various atmospheric levels as well as at surface and sub-surface levels. None of the 2,476 nuclear blasts around the world have ever produced an "EMP" that has eviscerated the electrical infrastructure in the areas they were detonated.

Magnetic Conjugate Regions Haven't Received Heightened Damage

A main plank of nuclear EMP theory is that "magnetic conjugate regions" and other areas on Earth with specific magnetic fields should receive even more damage than other regions. Tests have been made in those regions during high-altitude blasts and nothing significant was found by researchers or observed by populations living and working in those areas.

Cyclotron Radiation Frequency Never Detected

The EM cyclotron frequency was never detected or reported. Its existence is essential if the "deep" fully articulated EMP theory is real. If they really want to claim that they can find all the ingredients of EM electrical inductance from a nuclear blast (like flux at a low enough frequency to defeat the plasma frequency objection for metal conductors), then they have to have a coherent EM cyclotron frequency. No one will put the crony contractors' feet to the fire and press them on this point, but it is essential if they want their fancy theory to be consistent. The cyclotron frequency emanating after the blast should have been the biggest EM emission on Earth if it could fry electronics for thousands of miles in every direction. But, no. The supposed gigantic EM cyclotron emission was never observed by anyone: not by ham radio operators, EMP scientists, radio astronomers, military researchers—no one.

Why No Free Cyclotron Power From Solar and Cosmic EM Radiation?

If the deep EMP theory which espouses cyclotron radiation as the silver bullet is true, then we should be receiving directional electrical energy right now in wires from normal solar and cosmic radiation that would be converted naturally to cyclotron radiation at a lower useable EM frequency; useable from the perspective of inducing directional electromotive forces in wires on Earth. This doesn't happen. It is therefore clear that postulated cyclotron EM waves after a nuclear blast cannot feasibly be used as a way to forecast electrical flow in wires on Earth.

Overestimation of Cyclotron Effect

The cyclotron effect, if it happens after a nuclear blast, is very small because most free electrons bond with atoms in the atmosphere after the blast and don't spin down magnetic lines towards Earth producing unified EM waves in sync with all their fellows.

Exaggerated Intensity

Exponential reductions in intensity are not carried to their logical conclusion when researchers calculate the ultimate fate of all the energy emanating from a nuclear blast in outer space. As the sphere of effects around the blast expands, there is an exponential reduction in energy density in each portion of the sphere. An example of this exaggeration is seen when the ludicrous claim is made that EMP effects from a single blast could knock out power in the whole continental U.S. Also, a proper reduction is not made for most of the blast effects which go upward and outward into space along the path of least resistance. This process produces high-altitude artificial radiation belts around the Earth which carry a lot of the energy from the blast. This major portion of energy left in space should be subtracted from any that is included in grossly exaggerated EMP calculations.

EMP Experts: Nuclear Devices are Too Small or Too Big

It is hilarious to see the hijinks that EMP researchers will resort to when their theories fail to ever produce results. They switch back and forth between saying that kiloton blasts are too small and megaton blasts are too big. Yet, there have been multiples of every size blast in-between. Ultimately, it turns into some pointless discussion about pre-ionization to explain why an EMP is never seen.

Jokes

Wow, this section of the official study is great! I think Sandia put this section in their study because they were guilt-ridden due to all the empire-supporting lies they were telling. This section speaks for itself when it says that "EMP is not real." It is a real opportunity to see into the minds of the researchers and to see what they really thought behind all the gobbledygook

All the Proof Is Classified?

Another point to debunk is that EMP is actually real, but that all the proof of its existence is classified. Any massive infrastructure damage in Kazakhstan, on Pacific Islands, or anywhere else would be public

175

knowledge by now. The fact that it has never come out means that it never happened. There are plenty of residents of Kazakhstan that have talked about the blasts over their homes, but they don't recount living in a dark age from the destruction of the electrical infrastructure of the whole region. They don't have memories of that because it didn't happen.

EMP Producing Machines

Devices like the Z Machine and Trestle Machine do not simulate nuclear EMP effects because they employ totally different science as that attributed to nuclear EMPs. The machines use large instantaneous electrical releases. EMP advocates fail to mention that this is not how nuclear EMPs are alleged to function. Nuclear EMPs (per their own theories) allegedly come from outer space blasts that impact the atmosphere with gamma radiation producing a supposed cascade of events resulting in EM waves that induce an electrical flow in wires on Earth. These machines prove nothing.

30

Solar Flares

Same Factors Apply to Solar Flares

Incidentally, EM induction factors outlined in this book apply equally to solar flares—both the realities and the fictions. A solar flare would need to present a fluxing EM source that was at a low enough frequency to overcome the plasma frequency of the metal conductors in electrical infrastructure, just as would a nuclear "EMP." It would need a way to impart "directionality" on the circuit to cause electrical flow and produce damage. Most confirmed solar flare effects on human electrical systems are related to communications systems and come from the changed reflection and absorption characteristics of the Earth's ionosphere when impacted by solar energy.

Low frequency AM radio signals propagate long distances at night and short distances in the day, due to the sun's energy hitting the ionosphere in the daytime hemisphere causing absorption of radio waves. Other communications frequencies are enhanced by the solar energy. This daily solar cycle tends to affect radio, TV, and other EM communication signals traversing the atmosphere. In the same manner, propagation for some communications signals is enhanced and for others is diminished with varying levels of ionization from solar flares that either increase or decrease the absorption or reflection characteristics of different portions of the ionosphere. For example, radio operators are able to communicate over much longer distances—thousands of miles—on certain high frequency bands during peak solar flare activity due to the increased reflectivity of certain layers of the ionosphere. They actually look forward to large solar flares at the peak of the 11-year solar cycle. For them, it is not a hindrance, but a benefit.

When it comes to electrical flow in wires from solar flares, the same physics applies as mentioned previously in this book. Unlike a nuclear blast, the effects of a solar flare would impact the Earth's atmosphere and magnetosphere from a consistent angle and persist over a longer period of time. The amount of energy impacting the atmosphere and magnetosphere would also be much larger than that coming from a nuclear device and something like cyclotron radiation from individual electrons spiraling along the Earth's magnetic lines could come into play. However, we would still have doubts about the existence of consistent unified wave flux in the photons coming from the cyclotron effect. The action would have to produce something like consistent frequencies and not a lot of self-cancelling overlapping emissions if there were to be coherent separate photon waves pushing electrons through wires. Another problem for EM electrical induction would be the lack of appropriate changing angles from the EM source allowing it to push electrons through the entire wire circuit. And, of course, the energy coming from any cyclotron radiation would be vastly less than the initial solar energy impacting the atmosphere.

The same theoretical arguments for nuclear EMP damage are made to prompt panic from solar flares. The purpose of this book is not to cover solar flares, but there is a lot of similar speculative fear-mongering with solar flares. Theories are all over the place, just as they are with "EMP." Some say that solar flares naturally produce every EM frequency on their own from low radio frequencies up to gamma frequencies. How they could do this with no separate wave differentiation at every frequency is unexplained. Others would rely on a magnetobremsstrahlung / cyclotron effect to get the needed EM frequencies to induce electrical forces in wires. Others only talk about ionization effects on the ionosphere which have been realistically verified.

Whatever the theory, human electrical infrastructure has not been wiped out from solar flares. The Earth has had large electrified communities using long wire systems for over a hundred years. There is a peak in solar flare activity every 11 years. This is a normal predictable occurrence. We have survived 10 such solar peaks since there has been significant electrification on Earth.

Crisis advocates, of course, try to collect anecdotes of electrical "grid" failures and other incidents occurring on Earth during solar flares as they do during nuclear events, whereas they ignore them otherwise; anecdotes like the suggestion that a telegraph operator felt a shock during a solar flare in 1859, or that "safety systems" automatically shut off a power plant in Canada during a solar flare in 1989, etc. But, a wide area electrical induction effect on long wires or the zapping of small electronics has not occurred.

That should be the end of it. We shouldn't spend our time worrying whether some space Godzilla will breathe fire and blow out our electrical lines, digital watches, and pocket calculators. The crisis mongers want you to be fearful of such things and to fork over your money to be protected. Fight back by laughing in their faces!

Side Note: Local Outage

After working on this section of the book, I took a break and read the local news headlines that I receive in a daily email from a TV station (KGUN) in Tucson. Since my mind has been focused on electrical failures and their causes for this book, a story from the previous night (December 2, 2017) jumped out at me. It was titled "Tucson Electric Power Investigating Power Outages across Tucson Saturday Night." The subtitle of the article was "Thousands of Tucsonans Without Power Saturday Night." It then showed a map that had six areas highlighted where the power outages occurred. They were scattered across town. The short piece ended with this statement, "The cause of these outages is being investigated by Tucson Electric Power."

As I delved deeper into the local story as reported by various other local websites, I noted the six areas where outages occurred that night: Two in extreme Northwest Tucson (Oro Valley), two in Catalina (North Tucson), and two in South Tucson. The various stories said that the cause was "under investigation." The story held no special prominence, so it just faded away after a few days. I read one article about a car running into a power pole that weekend, but

179

that was never shown to be the cause and wouldn't explain the widely divergent geographic areas of the failures that were separated by areas that did have power.

Now, if this had occurred near the time of a nuclear test in the local area, it almost certainly would be attributed to a nuclear EMP. Even if the test had been in Timbuktu, it would conceivably be attributed to a nuclear EMP with stories saying something like "The EMP was carried along the Earth's magnetic lines causing damage in places as far away as Tucson, Arizona."

Also, if this had occurred during a large solar flare or during the peak of the 11-year solar cycle, we would very likely be seeing headlines on the national news like, "Solar Flare Shuts down Power Grid in Tucson."

More "scientific" news portals would no doubt add details and say something like, "Internal safety systems kicked in and automatically shut down the affected portions of the grid that were impacted by parasitic electrical flows resulting from the extreme solar event. Power lines have long been known to be magnets for dangerous electromagnetic currents coming from solar flares and nuclear EMPs."

As it is, we are well into the receding phase of the current 11-year cycle. The current cycle began in 2008. We are years past the peak of the cycle which was in 2013/2014 and are approaching the solar minimum which will be here in a year or two. So, this is a non-story. This everyday failure, and ones like it, will be forgotten. They will never be dredged up and repeated ad nauseum because they don't bolster a crisis theory. They happen all the time, but nobody cares.

With further reading, I have discovered that failures are so common in the Tucson area that the power company has dedicated a web page and features in a smartphone app to reporting and tracking outages. And, that is for a power company that only covers, at most, 1/300 of the population of the United States. I am sure that other utility companies have similar reporting and tracking systems for their very common outages. The point is that these outages are common yet forgettable when they don't coincide with nuclear events or peak solar flare activity.

31

What Can You Do?

"But, I'm Still Scared"

If you are still scared of EMPs, despite there being absolutely no indication that they are a threat (as a matter of fact, they have been proven empirically after 2,476 nuclear blasts to not be a threat), what can you do?

Well, the main thing you could do is to have a system that is not connected to long wires (phone, power, cable TV, etc.). Long wires are the alleged threat. Although long wires cannot receive an induced electromotive force from random scattering photons that are not in coherent waves, that is the allegation. Although long wire circuits would have self-cancelling forces on the back side if swept by EM waves, you are still supposed to fear.

So, if you believe the allegations, or want to protect yourself "just in case" the EMP malarkey turns out to be true, then connect your power to a local solar panel array or a local generator. Or, use battery powered devices. Then, you will not be at risk of drawing in the (fake) EMP electrical forces that would allegedly be induced parasitically in long-wire systems.

Although, it is a du jour fad to say that small consumer electronics would be damaged by an EMP pulse, this has never even been a serious allegation within the scientific community studying EMPs. So, do you need to put your consumer electronics or communications gear in a Faraday cage, a metal box, or under a tin roof? No, but if you were a ham radio operator (like me) and have long wire HF antennas strung out across your property, you may decide to disconnect your long wire antennas for the long wavelength radio spectrum below the 15 meter (21 MHz) band when not in use. These antenna wires are long enough to fit the profile of a parasitic long wire system that could allegedly receive induced nuclear EMP signals. Disconnecting any such long wire systems when not in use is another thing you could do if you were scared by the crisis-advocates.

32

Wrap-Up

This book has mainly focused on the grossly exaggerated "Hawaiian Street Light Incident," which has become the touchstone event for EMP researchers. Citing the incident as "proof of concept" is a rite of passage for EMP novices. It is clear, however, that the Hawaiian incident was a statistical non-event and that the research to prove an EMP cause was an exercise in pure fiction. The lighting system was shoddy and prone to failures. That's what likely caused failures on any night including the night when everyone was looking for failures. All records had been destroyed by the time of the EMP study, so there was nothing that researchers could use as a basis to form EMP conclusions. The fact that they proclaimed an EMP cause anyway is to their eternal discredit.

As shown in the previous sections, there are significant arguments against the speculative claims of induced voltages / currents in long-wire electrical infrastructure from a high altitude nuclear blast. It has never been shown that photons emanating from nuclear blasts arrive in coherent organized "fluxing" waves at an appropriately low frequency to effectively induce an electromotive force in accordance with Faraday's Law of Electromagnetic Induction.

Electricity through the Air

Such a thing as a single non-fluxing "EM" pulse simply cannot induce an electrical flow in long wires through a significant air gap of hundreds or thousands of miles, much less a flow that is capable of doing significant work (e.g. blowing fuses) at the recipient end. We are not talking about lightning or actual electrical flows arcing through the air.

Nikola Tesla tried to accomplish the transmission of electrical power through significant air gaps his whole life, but was unable to do so even when operating with a correct theory of fluxing EM waves as a source for the power (which is lacking with a nuclear EMP). The transmission of significant amounts of electrical power via EM waves over long distances through the atmosphere has never been demonstrated and remains, at this point, pure science fiction. The main obstacle is the inability to continually re-align a fluxing EM source with all the angles in the recipient circuit to make a consistent push of electrons through the whole circuit at a long distance. Even then, the EM source would have to be enormously powerful to have a slight effect on the recipient end after accounting for dissipation.

No Observed EMPs

The scare mongers say that a single electromagnetic pulse from a nuclear blast would induce a voltage spike that would knock out the "grid" and all electronic equipment across huge areas, like the entire continental United States. Something like this has never happened despite thousands of nuclear blasts and it does not fit with the established science of electromagnetic induction.

Isn't this really what matters? After thousands of detonations, an "EMP" has never swept through and wiped out electrical infrastructure, whether it was close to or far from a nuclear blast. Do we really need to say more? They used to say that your head would come off your body if you travelled at a speed of 60 miles per hour. They don't say that anymore. We can stop saying that EMPs are the boogey man that will destroy all electrical infrastructure. It's not a valid theory anymore. It never was. It has been tried and found wanting. There have been lots of nuclear blasts. It just doesn't happen.

33

Why I Wrote This Book

As with all things that Leviathan gets involved in, our freedom declines. There is always an inevitable tax impact as well, either direct or hidden through inflation. Invented crisis is always used to peddle the need for more funds and more hegemony. Fear-mongering helps gin up war tensions as well.

Sometimes an unwarranted big action, like a war, is initiated which turns into a desirable perpetual crisis factory—from the state's perspective. Never mind if we should have gotten involved in the first place. We are there and there is danger all around our troops, so we must stay. More money, more weapons, more power for the oligarchs. Never mind if the danger was manufactured or imaginary. They can always allege that fearful things are threatening us.

In the case of nuclear weapons, it is largely the same. Sure, the governments have the keys to their use and we don't trust the governments, but we shouldn't accept all the additional fear that automatically surrounds big state fear factories. The state is gleeful that within a war, within the nuclear arms world, within climate science, within public school, or within any other area of government involvement, there are always other peripheral scare tactics that can be deployed. There are always new fake things to be scared of that only the government can deal with. Things like EMP. They want to fan the flames and keep us in a fear frenzy. We are supposed to fear other people and fear a free society.

They are probably happy that there has been a moratorium on nuclear detonations. If test detonations in the sky had continued year after year, the EMP theory would likely be officially discredited and completely dead by now.

EMP Cult

The way it is now, EMP science is sort of an inherited magical cryptic belief system that is bequeathed to novices by academic forefathers, replete with imbedded signs and symbols after a profession of faith. A lay or professional EMP advocate can behave like a sorcerer, but instead of using chicken blood, he can sprinkle pointless calculations void of reality all over the place.

A person may pursue EMP in the same manner that others may invest time speculating whether Hitler made it out of the bunker and lived out his life in South America. Since Hitler's natural lifespan would be over by now and the high-altitude blast era is over, they can say about anything they want and have nothing to refute them. Actually, Hitler researchers would have more of a chance of turning up something real since they are dealing with something that, although statistically very unlikely, could possibly have happened. High altitude nuclear EMP is impossible.

We should laugh at them and their ridiculous notions. That is why I wrote this book. EMP is a ridiculous notion; one that we are supposed to give up our money, our common sense, and our freedom to validate. From the state's perspective, there is always some area of life where people haven't yet developed the proper level of panic to make them tolerate the forced filling of state coffers in relation to that area. There is always something new to fear that the public can't quite grasp without the government to ratchet up its fears.

Henry Hazlitt advised in "Thinking as a Science" that we "get a specialty" and that we "immediately put the idea or solution in writing" when we have "solved a problem."

What I have debunked in this book doesn't amount to much in the realm of problem solving because the flaws on the other side are so obvious and so numerous. But, either way, it is something that no one else may bother to invest time in or take sides on other than numerous paid cronies, state-lauding academia, or science fiction writers that benefit from crisis mongering.

Ron Paul on Government Commissions

To sum up this book, which mainly focuses on a government funded study of an event, I will conclude with something Ron Paul said about government commissions during an interview with Tom Woods on October 25, 2017:

"A commission is never designed to get to the truth. It's always to distort the truth and protect the government."

Appendix

This appendix shows a small portion of the ridiculous calculations that were included in the Sandia study. The calculations were supposed to lend an appearance of legitimacy to the unfounded conclusion that Hawaiian street lights were burned out by an EMP emanating from the Starfish Prime nuclear blast in 1962. Remember, these calculations were made using what Sandia admitted was "imaginary" data—since there were no records, wiring, or circuits to examine when the EMP study was done in 1989. Also recall that Sandia said, "Detailed calculations are not justified because of such uncertainties." That didn't stop them. They went ahead and produced overly "detailed calculations" designed to make the common man retreat from the subject with a sense of resignation due to his inability to understand. The calculations, however, are pointless without real data. They are merely window dressing on an atrocious work of fiction.

$$E_i e^{i\mathbf{k}\cdot\mathbf{x}} = E_i \exp[ikz \cos \psi \cos \phi - ik(x-h) \sin \psi$$

$$-iky \cos \psi \sin \phi] .$$

$$E_x(h,z) = E_i\, e^{i\mathbf{k}\cdot\mathbf{x}}\, (1 + R_V\, e^{i\omega 2h \sin \psi/c})\, \cos \psi \; ,$$

$$E_y(h,z) = -E_i\, e^{i\mathbf{k}\cdot\mathbf{x}}\, (1 - R_V\, e^{i\omega 2h \sin \psi/c})\, \sin \psi \sin \phi$$

$$\acute{E}_y(h,z) = -E_i\, e^{i\mathbf{k}\cdot\mathbf{x}}\, (1 + R_H\, e^{i\omega 2h \sin \psi/c})\, \cos \phi$$

$$E_z(h,z) = -E_i\, e^{i\mathbf{k}\cdot\mathbf{x}}\, (1 + R_H\, e^{i\omega 2h \sin \psi/c})\, \sin \phi$$

$$V(0) = -E_y^{inc}(h,0,0)\, d + 2Z_0 P(0)$$

$$+ \frac{E_y^{inc}(h,0,-p)\, d}{\cos kp} + \frac{Z_0\, e^{ikp}}{\cos kp}\, [Q(-p) - P(0)]$$

$$V(0) = -E_y^{inc}(h,0,0)\, d + \frac{A_1(\psi,\phi)d\, E^{inc}(h,0,0)}{1 - \cos \psi \cos \phi}$$

$$\times [1 - e^{ikp(1 - \cos\psi \, \cos\phi)}]\, V(0) = E^{inc}(h,0,0)\, d\, A(\psi,\phi)$$

$$V_{inc} = 10\ kV \rho_i = (3Z_0 - Z_0)/(3Z_0 + Z_0) = +1/2.$$

191

$$A(\psi,\phi) = \begin{cases} \dfrac{\sin\psi\,\sin\phi}{1-\cos\psi\,\cos\phi} \\[2em] \dfrac{\cos\psi+\cos\phi}{1-\cos\psi\,\cos\phi} \end{cases}$$

$A(\psi,\phi)$ (at $\phi = -9.8°$) $V_{oc} = 67\ E_{ih}(t)\ d - E_{iv}(t)\ d$
Z_o is $120\ \ell n(d/a)\ \Omega$. With $d = 0.368$ m and $a = 0.002058$ m, $Z_o = 620\ \Omega$

$$I_c(z,\omega) = \frac{4E^i(\omega)\ e^{-jkz\cos\theta}}{k\sin\theta\ Z_o\ H_o^{(2)}(kR_o\sin\theta)}$$

$$I_c(z,\omega) = \frac{-2\pi c\ E^i(\omega)\ e^{-jkz\cos\theta}}{Z_o\sin\theta\ j\omega\ \ell n(kR_o\sin\theta)}$$

$$I_c(z,\omega) = 2.7\times10^6\ \frac{E^i(\omega)}{j\omega}\ e^{-jkz\cos\theta}$$

$$I_c(z,t) = 2.7\times10^6\int_0^t E^i\left(t' - \frac{z\cos\theta}{c}\right)dt'$$

$$e_B = e_r \times e_{Einc} = 0.429\ e_E - 0.681\ e_N + 0.594\ e_V$$

$$C = \frac{\pi\epsilon_o}{\cosh^{-1}(d/2a)}\quad L_{external} = \frac{\mu_o}{\pi}\cosh^{-1}(d/2a)$$

$$V(z) = -\int_0^d E_y(h,y,z)\ dy$$

$$E_y = \frac{i\omega}{K^2}\frac{\partial B_z}{\partial x} - \frac{i\omega}{k^2}\frac{\partial B_z}{\partial x}\quad k = \omega/c \text{ and } c = (\mu_o\epsilon_o)^{-1/2}$$

$$V(z) = -\frac{i\omega}{k^2}\frac{\partial}{\partial z}\left[\int_0^d B_x^{inc}(h,y,z)\ dy + \int_0^d B_x^{sc}(h,y,z)\ dy\right] + \frac{i\omega}{k^2}\frac{\partial}{\partial x}\int_0^d B_z^{inc}\ dy$$

$$V(z) = c\cos\psi\,\cos\phi\int_0^d B_x^{inc}\ dy - \frac{i\omega L}{k^2}\frac{\partial I}{\partial z} + c\sin\psi\int_0^d B_z^{inc}\ dy$$

$$\frac{\partial I}{\partial z} + YV(z) = Y\left[\cos\psi\,\cos\phi\int_0^d c\,B_x^{inc}\ dy + \sin\psi\int_0^d c\,B_z^{inc}\ dy\right]$$

www.ingramcontent.com/pod-product-compliance
Lightning Source LLC
Chambersburg PA
CBHW071301220526
45468CB00001B/228